The Garden Retreat in Asia and Europe

The Garden Retreat in Asia and Europe

Ways of Dwelling in a Torn World

Edited by
Yue Zhuang, Alasdair Forbes and Michael Charlesworth

BLOOMSBURY VISUAL ARTS
LONDON • NEW YORK • OXFORD • NEW DELHI • SYDNEY

BLOOMSBURY VISUAL ARTS
Bloomsbury Publishing Plc
50 Bedford Square, London, WC1B 3DP, UK
1385 Broadway, New York, NY 10018, USA
29 Earlsfort Terrace, Dublin 2, Ireland

BLOOMSBURY, BLOOMSBURY VISUAL ARTS and the Diana logo are trademarks of Bloomsbury Publishing Plc

First published in Great Britain 2025

Cover design: Eleanor Rose
Cover image: A detail from Living Aloft: Master Liu's Retreat, 1543, Wen Zhengming (1470–1559), Metropolitan Museum of Art

A catalogue record for this book is available from the British Library.

A catalogue record for this book is available from the Library of Congress.

ISBN: HB: 978-1-3504-4738-7
ePDF: 978-1-3504-4739-4
eBook: 978-1-3504-4740-0

Typeset by RefineCatch Limited, Bungay, Suffolk
Printed and bound in Great Britain

To find out more about our authors and books visit www.bloomsbury.com and sign up for our newsletters.

CONTENTS

Part III Moral virtues and poetic virtues

Part IV The garden retreat and mortality

ILLUSTRATIONS

Plates

Figures

CONTRIBUTORS

Kasie Alt is Assistant Professor of Art History in the Betty Foy Sanders Department of Art at Georgia Southern University. Her research focuses on eighteenth- and nineteenth-century visual culture, including landscape design and representation. Her work has been published in peer-reviewed journals such as *Garden History*, and presented at various national and international conferences, including College Arts Association (CAA), SECAC, and the American Society of Eighteenth-Century Studies (ASECS).

Michael Charlesworth is Emeritus Professor of Art History at the University of Texas at Austin. Specializing in interdisciplinary approaches, he has in recent years written the first full-length study of the early twentieth-century plant collector, writer, water-colour painter, gardener and Buddhist, Reginald Farrer; and a critical life of the late twentieth-century film-maker, painter, writer, designer and gardener, Derek Jarman. He has published scholarly articles on early photography, the Picturesque and eighteenth-century panoramic drawing, as well as articles on the gardens of Stourhead, Rievaulx Terrace and Wentworth Castle. His interdisciplinary study, *Landscape and Vision in Nineteenth-Century Britain and France* was published by Ashgate in 2008. During the last two years, he has been occupied in writing a second book project about Derek Jarman, while living in a small wooden house in Austin, Texas, built in 1917, with a small garden round it.

David E. Cooper is Emeritus Professor of Philosophy at Durham University, and has been a Visiting Professor at universities in the USA, Canada, Malta, South Africa and Sri Lanka. He is the author of many books including *World Philosophies: An Historical Introduction*, *The Measure of Things: Humanism, Humility and Mystery*, *Convergence with Nature: A Daoist Perspective*, *Animals and Misanthropy* and, most recently, *Pessimism, Quietism and Refuge in Nature*. He is a regular contributor to the *Times Literary Supplement* and the *Los Angeles Review of Books*, and is also the author of several works of fiction. He has written widely on gardens, including his book, *A Philosophy of Gardens*, and several articles on East Asian traditions of gardening.

Born in London, **Alasdair Forbes** graduated with an honours history degree from McGill University, Montreal, before completing an MA in the history of art at the Courtauld Institute, London. He has worked in Paris and Vancouver, and lived for short periods in India and Italy. Before moving to Devon, he lived and taught in London. For the last thirty years he has been preoccupied with creating and looking after his garden. He currently teaches a course on gardens as places of meaning with Yue Zhuang at the University of Exeter. He has written a book on his garden: *On Psyche's Lawn: The Gardens at Plaz Metaxu*.

Stanislaus Fung is Adjunct Associate Professor of Architecture in the Chinese University of Hong Kong. His research work is focused on Chinese gardens and on contemporary Chinese architecture. Among his publications are essays in *Recovering Landscape* (edited by James Corner, Princeton Architectural Press, 1999) and *Thinking the Contemporary Landscape* (edited by Christophe Girot and Dora Imhof, Princeton Architectural Press, 2016). An account of his views on doctoral education can be found in *InterVIEWS: Insights and Introspection on Doctoral Research in Architecture* (edited by Federica Goffi, Routledge, 2021).

Ian James Kidd is an Associate Professor in Philosophy at the University of Nottingham. He has research interests in Buddhist and classical Chinese philosophies, including their attitudes towards gardens, animals, and the natural world. Much of this work reflects a wider interest in the topic of misanthropy, about which he is writing a book.

Iván Szántó is Chair of the Department of Iranian Studies of the Eötvös Loránd University (ELTE), Budapest, Hungary. Between 2010 and 2017 he was leader of a project at the Institute of Iranian Studies, Vienna, Austria, aiming to contextualize Islamic, chiefly Persian, art objects in Central Europe. In 2018 he was scholar in residence at the Getty Research Institute in Los Angeles, USA. His research interest covers diverse aspects of Iranian art history in the Islamic period, as well as Islamic artistic interactions with neighbouring regions. His books include *Safavid Art and Hungary* (Piliscsaba, 2010), *Artisans at the Crossroads: Persian Arts of the Qajar Period* (Museum of Applied Arts, Budapest, 2010) and *Honar-e Iran asr-e Qajar dar majmu'eha-ye Majarestan* (Tehran, 2016).

Minna Törmä is Senior Lecturer in History of Art at the University of Glasgow. She studied art history and theatre history at University of Helsinki and received her PhD with a dissertation on Northern Song landscape painting (*Landscape Experience as Visual Narrative*) in 2002. She complemented her studies with courses on Chinese art at University of California, Berkeley (1993–4) and the School of Oriental and African Studies, London (1998). She was Lecturer in the Arts of China (MLitt) programme at Christie's Education, London (2009–14). She is also Adjunct Professor of Art History at University of Helsinki. Among her many books and articles are *Enchanted by Lohans: Osvald Sirén's Journey into Chinese Art* (Hong Kong University Press, 2013) and *Nordic Private Collections of Chinese Objects* (Routledge, 2020).

Hongde Wu is Research Assistant Professor in the School of Design, Shanghai Jiao Tong University. His PhD thesis on the historic precinct of Tiger Hill in Suzhou argues that Chinese placescapes were created without modern presumptions of space and perspective. Aspects of his research have been presented in lectures and seminars in the Chinese University of Hong Kong, University of Hong Kong, Zhejiang University and Harvard University. His recent work has appeared in Chinese journals such as *Shidai jianzhu* (*Time + Architecture*) and *Jianzhu yichan* (*Architectural Heritage*). A book chapter entitled 'Sapience and Salience: Four Remarks on Beautiful China' (co-authored with Stanislaus Fung) has been published in *Beautiful China: Reflections on Landscape Architecture in Contemporary China*, ed. Richard J. Weller and Tatum L. Hands (ORO Editions, 2020).

Lin Zhu received her training in architecture (BA and MA) at Tsinghua University and worked as an architect for two years before starting her doctoral study at the University of Edinburgh in 2016. In Edinburgh, she investigated the literati's gardens during the Ming-Qing transition in the seventeenth century. Currently she is a research associate in the School of Architecture, Southeast University, China, working on a project on how Chinese gardens were made.

Yue Zhuang is Senior Lecturer in Chinese, Art History and Visual Culture at the University of Exeter. Her fascination with gardens began as an undergraduate studying architecture in China, and later guided her into an interdisciplinary enquiry into the history and philosophy of gardens and designed landscapes at Edinburgh and Zürich. She is the recipient of prestigious research awards from EU Marie Skłodowska-Curie Actions and the Leverhulme Trust. Exploring the intersections of landscape and garden traditions in China and Europe, her publications include *The Hermeneutical Tradition of Chinese Gardens* (Tianjin, 2015) and *Entangled Landscapes: Early Modern China and Europe* (Singapore, 2017).

ACKNOWLEDGEMENTS

Our first thanks are to all the authors for their contributions and their patient commitment to this volume. We also extend our thanks to the Department of Languages, Cultures, and Visual Studies (LCVS) and the College of Humanities at the University of Exeter for providing initial funds to organize an international workshop, 'Cultivating One's Garden: Eastern and Western Perspectives'. The workshop was held on 9–10 June 2022 at the Streatham campus of the university and the garden, Plaz Metaxu, Witheridge. The funds provided by LCVS also aided the production of this book, which has evolved from the workshop.

We would also like to acknowledge the students of the module 'The Place of Meaning: Gardens in Britain and China' at the University of Exeter for their enthusiasm and fantastic engagement, which was the main driving force behind the project.

We are grateful, as well, to the Architecture editor, James Thompson, the editorial assistant, Rosamunde O'Cleirigh, and the copy editor, Moira Eagling, at Bloomsbury for their professionalism and dedication to the book's completion.

Note on Chinese names and romanization

In the main text and in the notes, Chinese personal names are generally given in the traditional order, surnames first. There are a few exceptions: the Chinese names of the authors contributing to this volume are given in the Western order (surnames last). In addition, the names of authors such as Andong Lu are given in the Western order in accordance with the published form of their names in the works cited. Pinyin is used for romanization of Chinese names and words in the book. Romanizations in works cited are kept as they are.

Introduction

Michael Charlesworth, Alasdair Forbes and Yue Zhuang

In an experimental approach probing the possibilities of a new kind of garden history, this volume collects together a number of studies of gardens as works of art which respond to difficulties in the lives and world of the garden-makers. How to respond to a marred world; how gardens might become a resource in such responses; and how we understand the processes of living in gardens, are the ideas underlying such studies. We associate the phrase 'torn world' in our subtitle as first having a particular East Asian origin or bearing. The torn world suggests a rupture in the fabric of existence that goes beyond a profound dissatisfaction with the way people behave or the way things are run. The disturbance lies deeper, closer to the quick of what makes us suffer. In the famous eighth-century line of poetry from the Chinese Tang dynasty, poet Du Fu (712–70), '*Guo po shan he zai*' – 'the country is torn apart and rivers and mountains remain'– it is the devastating impact of warfare the poet has in mind. He contrasts this human violence which splinters lives with the steadfast integrity of the natural world. The Japanese Buddhist monk Kamo-no-Chomei (1155–1216) goes further. The torn world he conjures up in his poetic masterpiece *Hojoki: Visions of a Torn World* extends beyond a scathing indictment of human greed, belligerence and vanity to express a universal dismay at the violence in the natural world itself (he had witnessed first-hand a series of natural calamities: earthquakes, whirlwinds, fires, famines, floods).[1]

In addition to natural calamities, and outright warfare (as also impinging on the destruction of material records, or on the garden-maker's imagination, in Chapters 2 and 5), other factors can induce a profound conviction of the torn-ness, or unsatisfactory state, of the world. As Lin Zhu explains in Chapter 1, the fall of the Ming dynasty in China in 1644 weighed heavily on Qi Biaojia (1602–45), a government official, lover of theatre, and one of the

Chinese literati, with the most dramatic result. Violent political change can happen in the West, too, and Kasie Alt explores the ramifications of this in an example from nineteenth-century France in her Chapter 2 about a garden designed by one of the inventors of photography, the artist and professional stage designer L. J. M. Daguerre (1787–1851). Here is the one example in our volume of a professional designer working with the landowner to institute an already culturally established style of garden, as a result of his judgement about what design would best answer the needs of the moment in a garden intended to provide work for those who had lost their form of subsistence through the revolutionary upheavals of 1848.

So the world gets torn apart for people in the West, too, and in fact tornness is not specific to East or West, but has a planetary reach. As we are all too aware in the era of the COVID-19 pandemic, the effects of which are still being felt at the time of publication, the world can be 'torn' by disease; Michael Charlesworth (Chapter 10) explores how the English film-maker, artist, writer and stage designer Derek Jarman (1942–94), by relying, amongst other means, on the example of Far-Eastern *mandalas*, created a garden of commemoration of his friends who had died of AIDS-related illnesses. The idea of commemoration is pursued, in addition, through several Asian gardens by Iván Szántó in his engaging hunt for monuments to the famous Persian poet (and astrologer) Omar Khayyam (Chapter 4).

As can pandemic disease, armed conflict can also have the consequence of causing people to become pent up, confined. This happened during the Second World War to the eminent Swedish art historian and sinologist Osvald Sirén (1879–1966). Confined to home when he would have preferred to be travelling in his beloved China, Sirén lived on his memories, which together with his photographs helped him to create a garden of 'imagined space', as Minna Törmä shows in Chapter 5. Such a response to adversity links Sirén with an earlier Western traveller to China and Tibet, Reginald Farrer (1880–1920), whose life in England during the First World War after his return from an ecstatic experience of two years in remote northwest China and far-eastern Tibet resulted in his production of immensely readable memoirs of travel, and in the creation of a Chinese garden at his family home of Ingleborough Hall, Yorkshire.[2] For Farrer, happy memories were 'bottles of medicine', to be sipped whenever necessary. In Sirén's case, among their other effects, happy memories helped to prepare for the writing of his magisterial pioneering study, *China and the Gardens of Europe* (1950). This book took seriously the eighteenth-century French designation of the new landscape garden as *le jardin Anglo-Chinois* (the Anglo-Chinese garden) and Sirén enthusiastically found the influence of Chinese gardens in many European gardens, especially in their elaborate rockwork.[3]

Warfare (including civil conflict), natural calamities, political change, disease, and, in our own time, the planetary catastrophe of human-induced global overheating and other climate change, tear the world surrounding the garden-maker apart. To these factors we should add religious differences, whether or not these lead to the brutalities of warfare, terrorism and counter-terrorism. Overt brutalities on the battlefield or scaffold did not impinge too directly on the life of the English poet Alexander Pope (1688–1744) but, afflicted as he was by lifelong ill-health, he felt, and deeply resented, the impact of parliamentary discrimination against Roman Catholics that was virulent in Britain during his lifetime. Against this background, Yue Zhuang (Chapter 9) explores the spatial layout of his famous grotto and garden, and, with help from Pope's literary achievements, the thematic sequences suggested within those spaces by means of drawings by Pope's friend William Kent and his gardener John Searle, by garden furniture and constructed features, and by patterns of planting. The example of Alexander Pope can remind us of the way in which our worlds can be psychologically torn up. That is to say, how our stable, happy, optimistic view of a world full of promise can be destroyed by interpersonal tragedy, emotional trauma, loss of loved people, and the sudden onset of disability. Pope's example includes his handicapped body, but his garden shows how personal difficulties, explored through art, can in a compensatory way lead to a greater understanding of the human mind, of culture, of mortality and what we would now call the 'unconscious'.

In fact, reflection suggests that so manifold are the causes of the world being torn that we could well reach the conclusion that the world is always beset by disaster, or just unsatisfactory, and that therefore the expectation of a benign world becomes a delusion. Given that the torn-ness of the world is a property of both external reality (as grasped by personal observation and scientific research) and subjective dynamics and impressions, an unwrecked world remains an ideal.

Building on this, we might also notice how many of the chapters in this book tend towards the subject matter of death and the idea of mortality. The search for Omar Khayyam's (1048–1131) memorials, his grave, and the related gardens discussed by Iván Szántó; Qi Biaojia's suicide; Jarman's garden of commemoration; Pope's mind haunted by mortality: death seems to loom over the subjects of these studies. On one level this effect is surely related to our culture's suppression of the contemplation of death. As one of the features of the Scottish poet and artist Ian Hamilton Finlay's (1925–2006) garden at Little Sparta puts it in a counter-position, borrowing an inscription from the seventeenth-century poet Henry Vaughan, 'The contemplation of death is an obscure melancholy walk'; at Little Sparta this inscription is disposed between several plinths arranged along a garden walk overshadowed by trees; and as the poet completes the idea and the walk leads to a light open space: 'but it leads unto life'. In contrast to the dynamic extolled by Vaughan and Finlay, for too many of us, individual death, death of a civilization, and the harrowing possibility of the death of a planet can loom up and then be denied – the last for political reasons in a system dominated by knuckle-headed old-fashioned capitalism in which the exploitation of fossil fuels is central. The denial by our culture of these varied inflections of death leads to extraordinary problems. As the psychologist James Hillman, whose work is important for at least two of the garden-makers featured in this volume, Derek Jarman and Alasdair Forbes, put it as long ago as 1979: 'What is most dead and buried in each of us is the culture's neglect of Death. Hades only now begins to reappear in ominous new concerns with the limits to growth, the energy crisis, ecological pollution, ageing and dying.'[4] Hillman's view is that getting in touch with the idea of mortality is psychologically beneficial, not least because the important archetypal psychological forces, which he images as mythological figures, have a presence in our consciousness similar to that of the shades in the ancient Greek underworld presided over by Hades: shadows with no material or physical existence, but a potency that manifests in dream and the imagination as human-like figures.[5] On the level of civilization and planet, contemplation of death would help us to prepare for it, and even to prepare against it (for instance, on the ecological level).

Contemplating a 'torn world' may lead our thoughts towards mortality, but many people experience the frustrations and inadequacy of the world in terms of the human questions of how people behave or how things are run on a day-to-day basis – the very vexations that we brushed aside in the first paragraph of this Introduction. In these cases, just as in the cases of gardeners vividly aware of war, disease and catastrophe – and indeed, in contrast, with people who are satisfied by the world as it is – the garden can be a sanctum: somewhere relatively insulated from the economic and social pressures of the world outside. One example of the turn away from a vexing and frustrating professional life to private absorption in a protective sanctum (though with less dramatic consequences than in the case of Qi Biaojia) forms the basis of Hongde Wu and Stanislaus Fung's study of the sixteenth-century garden, Zhuo Zheng Yuan (the Garden of Artless Administration), made by Wang Xianchen in 1530s Suzhou, China (Chapter 3). This garden offers us a paradigm that is potentially repeated all over the world: on the death of his father, a government official retires, when still relatively young, after a rather chaotic career in order to make a complex work of art by allowing his artistic and personal integrity, constrained in his work life, to come to the fore. Here, as in the case of Pope's garden, crucial surviving evidence was provided in the form of paintings and poems, in this case made by an artist friend of the owner, Wen Zhengming (1470–1559). Wu and Fung study how the garden was represented in Wen's depictions and poetry with the theoretical help of contemporary philosophers of aesthetics, inducing a closer understanding of the artistic integration of emotion and the natural world.

In one section of this volume, two philosophers, David Cooper (Chapter 6) and Ian Kidd (Chapter 7), with the help of Western and – in Kidd's essay especially – Eastern traditions of thought, attempt to come to terms with gardens within the overall idea of refuge and the potential benefits to the human spirit a garden retreat can induce. By offering general thoughts, not tied to investigation of specific gardens, these short essays offer readers a different form of approach to issues that we hope to be present throughout the volume: gardens as means for the recentring of individual integrity, even in a moral sense, and even while surrounded by unpropitious external circumstances. Kidd mentions reverie as something for which a garden can be a resource, and in particular the example of the great French philosopher and novelist, Jean-Jacques Rousseau. For Rousseau in the eighteenth century, reverie itself was a precious path to the re-establishment, within the individual, of a sense of the creating deity's benign purposes. Rousseau experienced at least one of his reveries in the landscape garden of Ermenonville, some 30 miles northeast of Paris, where a feature in the form of a Roman altar is dedicated as a consequence of Rousseau's experience: 'To Reverie'. As a result of such ethical refreshment, the individual can emerge from reverie with a revolutionary conviction, based on the sense of fellow-feeling for other suffering human beings. Such ideas helped lead to the French Revolution, but can be still relevant in our own era.[6]

Especially crucial to seeing the garden as a sanctum, or refuge, is Alasdair Forbes's contribution to the present volume (Chapter 8). This supplies the dimension missing from the essays by Cooper and Kidd: a study of the value of a single garden to a specific person. It is extremely rare for students and scholars of garden history to find a detailed account of a (in this case fairly large) garden written by the garden's owner/maker. Forbes's essay may, moreover, be unique, and therefore invaluable, because it concentrates on describing the mental and emotional processes of the maker/owner that provided the energy for the endeavour. Another unique quality of this essay is that, except where necessary, Forbes avoids the gardener's usual material details of what species or varieties of plants were planted where. Instead of this dimension, he gives us an account less of material features than of the maker's sense of space in the creation of the garden; that is to say, the account is of a series of *spaces* that themselves make up the garden's 'features'.[7]

Refuge thus becomes a key term in the conception of the present volume, and of the workshop held at the University of Exeter in the early summer of 2022 out of which this volume stems. We have, however, decided to prefer the term 'retreat' in our title. Both words designate secure places of withdrawal from a too-active, even threatening, surrounding world. To this meaning, 'retreat' also supplies an additional psychological definition: it can mean a period of restorative seclusion, retirement, or solitude. Such an element is absent from the word 'refuge', in which psychological dynamics are not directly implied. Finally, a persuasive factor in the adoption of the word 'retreat' was provided by the example of Ian Hamilton Finlay. In his 'Unconnected Sentences on Gardening', modelled on the 'Unconnected Thoughts' of the eighteenth-century poet William Shenstone, Finlay famously wrote: 'Certain gardens are described as retreats when they are really attacks.'[8] Hinging everything on a pun, Finlay here plays upon an additional definition of the word retreat, the military withdrawal, to emphasize that a garden-maker can form, even in their own garden, satirical features that contest, correct or simply protest the delinquencies and hostilities of the surrounding culture. The French artist Paul Gauguin supplies an incisive example. During his activist phase in his home in the South Pacific islands, Gauguin refused to pay taxes to the colonial authorities, tried to defend native islanders taken to court for drunkenness, and attempted to persuade islanders not to send their daughters to colonial schools. In consequence he found himself attacked from the pulpit by the local Roman Catholic bishop. He responded by carving devilish sculptural caricatures of the bishop and displaying them in his garden, which lay beside the public road, so that any passer-by could see the satirical attack and smile.[9] In the present volume this polemological aspect is not particularly strident, but at least Qi Biaojia's, Alexander Pope's and Derek Jarman's use of their gardens qualify those places as retreats containing implicit reflections on surrounding socio-political circumstances.

In this respect, a balance has to be found between reclusiveness and visibility. In China, recluses were known as *yin shi* (hidden men). From Roman times, the ideal of retirement was linked with the desire to remain hidden, perhaps most famously in Horace's line *Secretum iter et fallentis semita vitae* (a secluded journey along the pathway of a life unnoticed), which Pope had inscribed above his grotto at the entrance to his garden.[10] The line is also quoted by Abraham Cowley (1618–67) in his influential essay *Of Obscurity* (1668), where he tells us that 'For my part, I think it is, and that the pleasantest of life, is in *incognito*'.[11] For scholar-poets in both China and Europe the garden became a prized setting for leading a life of relative secrecy. It is a rare hermit, though, who remains immune to the paediatrician and psychoanalyst D. W. Winnicott's observation that 'It is joy to be hidden but disaster not to be found.'[12]

Refuge or retreat becomes all the more important as a hermeneutic device and a theme because the gardens studied in this volume are, with one exception, not products of professional designers imposing conventional styles of design before moving on to do the same thing in other places. One of the rare values of this volume is that it deals with gardens made by the owners, or inhabitants, of the spaces. An owner/dweller puts together features that they want. This brings up another term from our title, which is 'dwelling'. The emphasis on dwelling diverts attention from professional designers to individuals who lived in their gardens (or in the case of Omar Khayyam, who foresaw his own grave). Such an emphasis also diminishes the importance, which has become rather conventional in garden history, of a succession of styles of gardens that supposedly happened as time marched on. Perhaps the example in which professionalization and style still figure most strongly as factors of agency is that of Daguerre in Bry, choosing the Picturesque style of landscape garden for someone else's domain and for social benefit as a comfortable nostalgia in a time of acute political crisis.

The torn-ness of that time was compounded some fifteen years after Daguerre's death when the Prussian invasion of France overwhelmed the town, there was fighting, and among other things all the domain's archival records were burned. The significance of Bry's garden, as Kasie Alt shows, has therefore to be reconstituted from other sources.

In the case of the garden-dwellers (*habitant-paysagistes* to use Bernard Lassus's formulation) motivation, planning and making are internally generated. The tenacity of such an impulse, even among the least fortunate members of society, is documented poignantly in Diana Balmori and Margaret Morton's study of gardens made by homeless people in New York.[13] In such cases, just as much as in the examples studied in the present volume, the word 'dwelling' connotes not just living, but lingering, even concentrating. To dwell *on* something means to give it considerable thought. The thought does not have to be particularly organized, to follow a chain of logic. It can revolve around its subject, while still giving the subject deep attention. As such it is perhaps close to Rousseau's reveries, and favours a poetic approach to a subject rather than a strictly rational one.

This volume presents some cases that offer paradigms of how to dwell in a torn world. Its strength is in the variety of approaches exemplified and employed. As a counter-balance to variety, though, many of the examples share an imaginal or ideational literary energy, by being closely associated with poetry. What were the motivations behind all this garden-making? At least for Qi Biaojia, Wang Xianchen, Sirén, Jarman and Forbes, we can specify that the motivation was love. Love of theatre, of art, love of Chinese paintings and gardens, love of nature, love of the challenges, and, for Jarman, love of other men. In the end, the common factor is love of gardens. We know from the work of the philosopher G. E. Moore that love sits within the highest ethics. Moore clarifies in his *Principia Ethica* how the highest ethical states we experience – the greatest ethical 'goods', as he calls them – are certain states of mind that we get into in connection with love, friendship and the contemplation of beautiful objects (which can be natural or man-made).[14] And here we have one of the resources and strengths of gardens as works of art, as they combine man-made and natural features into beautiful arrangements and wholes.

* * *

In this second half of the Introduction, we put forward, through the way in which the essays are grouped together, a suggested cross-pollination of themes that the essays inspire, some mutually reaffirming and some unexpected. The patterns thus emerging are neither exclusive nor definitive: the reader is invited to experiment with other possibilities of their own. In line with the 'experimental approach' named at the outset, we hope these cross-cultural and even a-historical juxtapositions will provoke a more open and flexible understanding of the cultural importance of gardens and the way they may be written about.

The garden retreat and theatricality

Our discussion opens by considering two gardens created during times of war and civil conflict. Despite drastic tempo-spatial differences (seventeenth-century China and nineteenth-century France), both gardens – Mountain Yu in Zhejiang Province and the park of Bry-sur-Marne in the eastern suburbs of Paris – engaged with theatricality and illusion to seek a way out of the deeply troubled reality.

In Chapter 1, Lin Zhu introduces a remarkable garden made at a time when the Ming dynasty was crumbling and the literati class resorted to 'garden mania'. Qi Biaojia, the owner and designer of the Mountain Yu garden, was an eminent scholar-official. Disillusioned by politics, he had several times resigned from, but been repeatedly called back to government. The garden retreat, for Qi and his contemporary literati, was not just a site for physical recreation and a place for fulfilling filial devotion.[15] Rooted in the imagination, the garden provided a mental and spiritual refuge.[16] This sense of refuge draws its power from theatricality and illusion, deliberately cultivated 'to transport the master from reality'. But this is not simple escapism, as Zhu argues. At the core of the garden activities or theatre performances (Qi was an excellent theatre critic) there was a profound contemplation of the fugitive nature of all existence and a transcendence of the boundaries between life and death. As Zhu emphasizes, not only was the garden deliberately constituted by the play between numerous opposed factors such as reality and illusion, waking and dreams, and day and night, but the garden thrived by 'presenting the unstable boundaries between these opposing elements' as the world we are invited to enter, a world whose oscillations we are encouraged to 'transit in between'.

In Chapter 2, Kasie Alt takes us to the park at Bry, designed and begun in 1848 by Louis-Jacques-Mandé Daguerre in the landscape of the château in Bry-sur-Marne. Daguerre, as Alt illustrates, conceived the park as a 'natural diorama' – a multimedia spectacle creating an immersive theatrical experience that blurs the boundary between illusion and reality. Commissioned by Mlle de Rigny, the owner of the château, the project was a response to her fear of socio-political upheaval following the end of the July Monarchy and the removal of the *Ateliers Nationaux* (national studios) as an emergency relief programme to employ citizens. Behind the landscape of the natural diorama, therefore, was the intention of avoiding potential violent revolts by occupying workers on Mlle de Rigny's property. The concept of a 'natural diorama' suggests the landscape park was created as a theatre of illusions which was simultaneously modelled on, and perfected by, a vision of art's ability to reveal nature.

Hinged on theatricality and illusion, conceived as a strategy for mitigating the crises, the links between the two gardens are remarkable. In Mountain Yu, Qi Biaojia, the lover of the theatre in his moonlit garden, was not estranged from or indifferent towards the collapsing Ming. Rather, as Sophie

Volpp has put it, the retreat to the garden and the culture of the theatre 'fostered a quality of perception that was simultaneously impassioned and dispassionate, that was immersed and participatory as well as cognizant of the positive value of disillusion'.[17] While the Confucianist call of 'letting life go and choosing righteousness' (*she sheng qu yi*)[18] was the fundamental reason for many Ming loyalists' suicide, the Buddhist view of illusion and disillusion, the Daoist idea of dream and wakefulness, which were intrinsic to the imagination of the Ming literati and embodied in the moonlit garden scenes, facilitated Qi's ultimate transcendence of the boundaries between life and death. Qi's ritual suicide in his garden is both an exemplary public moral gesture (a noble act of defiance against a corrupt – or barbarian – regime) and a final exalted vindication of a rehearsed equanimity in the face of death.

Given the paucity of documents left by both Daguerre and Mlle de Rigny, we may never know for sure whether the Diorama-like landscape park, 'an all-encompassing spectacle' that relied on the imagination, might have been conceived in alignment with earlier Baroque art – and the often overlooked religious dimension of the Picturesque[19] – as an instrument to demonstrate to the community of the faithful the inherent logic of the creation of the landscape as a work of God. Such speculation might be encouraged by the Diorama that Daguerre earlier installed for a local church in the village of Bry. As Alt demonstrates, the church Diorama created an illusionistic cathedral inside the church, 'a representation of the heavenly House of God'. For Daguerre and Mlle de Rigny who were devout Catholics, the Diorama would alert them to the Platonic-Christian dialectic of nature and art, truth and image (representation). Some visual evidence of Daguerre's grotto at the park further suggests religious allusions. The park, as a natural diorama, would have been regarded by the designer as an attempt to reveal nature's order through art rather than to make nature submit to art. Participating in the mass and the rituals central to the church, the illusionistic painting of the cathedral, Alt suggests, formed 'a liminal space that allows for a connection and crossing between physical and invisible realities'. One may wonder therefore whether the picturesque vision at Bry park, which imitated 'true nature', similarly constituted a revelatory exercise that exalted the beauty of nature designed by God over the reality of the moment.

Dwelling, garden and the poetic image

While war and civil conflict determines a torn world, the fact that each of us is 'thrown' into this world as a transient creature characterizes our being in the world as a poignant experience; in other words, the world which we as mortals are able to experience is necessarily a 'torn' one, a fragment of deep (non-linear) time and space as a continuum, an eternity.[20]

To dwell in the torn world, as we see in the second group of essays (Chapters 3–5), the poetic image – as embodied in garden, poetry and painting – becomes an ideal medium of repair for the rupture between the here-and-now as individually experienced and more communal and universal imaginaries of dwelling, which are embedded in, but ultimately transcendent of, differences of time, geography, nation or culture. This act of repair is cognitive and poetic; it refers our more or less quotidian existential unease to the wider horizons of what it means to dwell on earth, thereby relating our habitual sense of chronological time to deeper universal patterns.

In the three essays that make up this second section of the book, we encounter, first, a sixteenth-century album comprising intimately juxtaposed contemporary poems and paintings of a literati garden in Suzhou, the Garden of Artless Administration (Chapter 3); second, a sequence of gardens, extending from the Middle East to South Asia, each of which was inspired by the image of the garden generated by the posthumous fame of the verse of the eleventh-century polymath Omar Khayyam (Chapter 4); and finally, a garden made with the aid of hundreds of photographs of his travels in China and Japan by the Sinophile Swedish art historian, Osvald Sirén, on an island in the Stockholm archipelago during the Second World War (Chapter 5).

In all three cases, an intense assimilation of the garden and poetic image leads to a vision of the garden as a dwelling place in the way already outlined earlier. Being a gardener allows one to care for and nurture both growing things and crafted spaces where work, life and dreams unfold and interweave on a daily basis and where, to borrow Alberto Pérez-Gómez's words, the gardener learns to 'materialise the imaginary'.[21] All three gardens (or indeed, all case studies in the volume) are, like poetry, metaphorical structures; and metaphor, to stay with Pérez-Gómez, 'emerges from embodied consciousness oriented in the world, affording possibilities for communication among humans both in ordinary situations and in the extraordinary occasions when artistic artifacts speak out of time and place'.[22]

For such 'ordinary situations' and 'extraordinary occasions', we find examples in each of the three essays: thus, Wang Xianchen and Wen Zhengming, the two literati friends, sit in a wooden pavilion by a murmuring stream in the Garden of Artless Administration; or, the humble grave of Khayyam situated at the foot of a garden wall is covered by the petals of many blossoms and open to the sky – an image that would resonate for a millennium; or Sirén is moved to recreate in his own garden the atmosphere of reflections in a tranquil pond of a garden in Southeast China he had photographed many years before. All such images are indebted to specific garden vocabularies which variously convey the power of the dwelling-place.

The pre-eminence, indeed the indispensability, of the poetic image is especially characteristic of Chinese literati and Persianate gardens, as Hongde Wu, Stanislaus Fung and Iván Szántó show respectively. If the gardens materialize the imaginary, then it is the poems which invent that imaginary in the first place. As Gaston Bachelard suggests, the poetic image places us at the origin of speaking being.[23] Only reachable through language, the poetic image reverberates into the depths of existence; it gives rise to a psychic activity which resonates through its own kind of wavelength, putting us in contact with deep reservoirs of feeling and felt thought normally remaining inaccessible. Gardens, as these essays suggest, are, or can be, unrivalled as settings for tapping into these poetic reservoirs.

For Wang Xianchen, the owner and designer of the Garden of Artless Administration, and Wen Zhengming, the creator of the album of 'thirty-one scenes' of the same garden, garden and album combine to connect the two friends to the realm of the poetic image cherished by generations of educated Chinese recluses, who were inspired to cultivate inner enlightenment after becoming frustrated with life in public service. Through a close reading aided by the phenomenological analysis of three of the thirty-one scenes, Wu and Fung demonstrate that the garden album did not seek to establish any objective realization of the physical garden. Instead, the album, essentially an interplay between text (poetry) and image, may be understood as a milieu/medium that encompasses various perceptions, memories, and imaginations. Such a milieu/medium was primarily established by the semantic field of poetry – 'the extensive use of allusions and metaphors', of which Wu and Fung's essay provides ample illustrations. The painted scenes by Wen, similarly, maintain 'the sense of an open work'. Never intended to be a simple depiction of an actual scene, the paintings were capable of evoking multiple places and associations all at once, as if in a dreamscape. In the form of the album, the materiality of the garden is radically transformed and condensed into a poetically imagined plurality of worlds, an atmosphere, which enabled Wen and Wang to cultivate a 'peripatetic sense of moving between here and now and there and then'. In this free shuttling between reality and imaginative reverie, Wen and Wang's portrayal of the garden retreat provides us with an invaluable insight into how poetic dwelling was conceived and experienced by the Ming literati.

While Wu and Fung's chapter demonstrates the power invested in the poetic imagination by the Ming literati to transport them to an alternative realm of deep time and space synchronically, a similar poetic force is manifested diachronically in the gardens of the Persian-Arabic-Hindu region in Chapter 4. Here Iván Szántó takes us on a remarkable itinerary tracing the generative power of the poetic image of the garden as found in Omar Khayyam's poem, *Ruba'iyat*, initially through its apprenticeship abroad in Europe, but then through a series of inspired homecomings in Kathmandu, Kashmir and Kabul, back finally to Nishapur in Iran, where Omar Khayyam's tomb is located, and from where, in the form of a mystical encounter, the Khayyam-esque initiation into the poetic microcosm of gardens could be said to have begun. In Persia, as much as in China, poetry and gardens are synonymous with a national cultural identity. So much so, that by bringing a line or two of poetry into conjunction with a garden, a 'miniature Iran' can be created anywhere. While a garden provides sensory stimuli to the mind and heart, poetry that is capable of activating the material imagination has a greater capacity to recall the vivid nature of a 'metaphorical quintessence of Persia' which constitutes a shared commonality, carried in memory, in the mind and the heart. The improbable way in which Omar Khayyam's verse has transmitted the idea of the Persianate garden so successfully across both centuries and continents itself bears compelling witness to the power of the poetic image, and of poetry. Khayyam himself was known in his lifetime primarily as a scientist and an astrologer, gifted with powers of augury. In Balkh, in 1121, he had predicted, in front of his pupil Nizami Aruzi Samarqandi, how his grave would one day appear to him. Twenty-four years later, Nizami was astounded when, in Nishapur, he found himself beside his master's grave 'strewn with blossoms blown by the north wind' just as had been foretold. Szántó ends his itinerary in Nishapur, where a modern memorial has been erected for Omar Khayyam, after centuries of neglect, in large part because of the role played by the transmission of his garden poetics in the rehabilitation of his reputation.

Minna Törmä's Chapter 5 exemplifies a variant style of dwelling poetically in a garden through Osvald Sirén's cultivation of photographic images as a source of creative inspiration, profoundly influenced by his long-standing first-hand familiarity with East Asian gardens, art and culture. Having travelled in China and Japan for extensive periods as well as being well versed in Chinese painting and poetry, the Finnish-Swedish art historian Sirén was isolated in his home on Lidingö at the outbreak of the Second World War. Nevertheless he found a spiritual refuge in the imagined spaces of East Asia: the poetry of Du Fu and Li Bai (701–62), the painting of Guo Xi (1020–90) and Ma Yuan (1160–1225), the mountains of Huashan, as well as the many gardens through which he had wandered in China (Beijing and Jiangnan) and Japan. The hundreds of photographs that Sirén took during those travels, as Törmä tells us, were an aid to his memorial reflections, as were the pictures of paintings of China and Japan which he had accumulated through decades of research. The way Sirén engaged with his photographs is *wo you* (travelling while lying down), following the manner of ancient Chinese literati who enjoyed mind travel among streams and mountains painted in a landscape handscroll. Such mind travel in turn inspired Sirén's creativity – both in his own garden with echoes of East Asia (as in the rocky hill with a few twisted pines, or a Japanese pagoda) and in his literary productions such as *Gardens of China* and *China and Gardens of Europe in the Eighteenth Century*. Revealing 'unseen reality', Sirén's poetic imagery conversant with both Europe and East Asia may thus transport one out of the distressing circumstances of war and captivity into a redeeming haven in which the garden refuge played the central part.

Moral virtues and poetic virtues

The third group comprises three essays illustrating two different kinds of virtues available for exploration in a garden retreat. Chapters 6 and 7 by the philosophers David Cooper and Ian Kidd respect and vindicate the traditional values of the garden retreat or refuge as a sanctuary from the ills of the world, and a place where it becomes possible to develop moral virtues, to practise a form of self-cultivation.[24] The main reason for the garden to be the site for self-cultivation is, as Cooper suggests, its dependence on the natural world. This dependence ideally helps mark out the garden as a place to dwell, rather than as a mere alternative, or adjunctive, living space. Cooper's and Kidd's evaluation of the garden refuge is closely informed by the earliest philosophical appreciations of the value of the garden in both Europe and China as well as by modern philosophers such as Martin Heidegger. The preferred model of dwelling which their accounts of the garden present has a characteristically moral, rather than a poetic, emphasis (though the distinction is not an absolute one). The garden is where we can pursue the 'good life', not in the modern sense of a life of relaxed hedonism, but in the ancient sense of a life of *eudaimonia* (human flourishing) which understands happiness as, while open to multiple pleasures, inseparable from measured and moral reflection. It is through this form of self-cultivation that we may be able to dwell in the sense Heidegger articulates as 'to be set at peace', 'to spare and preserve things'.[25]

David Cooper writes in particular about garden practices, 'quiet' virtues, and understanding, virtues such as 'patience, humility, care, hope' and the 'respectful reverence' for things typified by the Chinese word *jing*. These 'quietist' virtues, Cooper argues, are 'internal' or 'intrinsic' to 'garden practices', and they foster an attitude to life that favours the 'happy, calm and unshaken' equanimity recommended alike by Greek and Chinese philosophers. Cooper encourages us to imagine the garden as a theatre set apart from the 'angry, agonistic world', where the forsaken – and even, today, despised – virtues of 'humility, forbearance and innocence' can be painstakingly remembered and allowed to flourish. Alongside the cultivation of these virtues ideally the garden will help encourage us to refine and enlighten our sensibility and to reach a wiser understanding of our place in the world. The natural setting of the garden makes it a distinctive kind of refuge, one unusually receptive to the 'mutual dependence – the reciprocal inflection and porousness' in play between the human and the natural worlds.

Ian Kidd in his chapter begins by emphasizing how the garden refuge removes itself from the 'corruption and failings of the mainstream world'. While originally the garden refuge kept the 'threats of nature' at bay, today it is essentially a 'refuge from men'. Kidd identifies three primary functions of the garden refuge, applicable both in Europe and China: it acts as a respite from the pressures of life; it restores 'damaged moral virtues and crumpled sensitivities'; it gives rise to reflection and reverie, which are valuable components of the good life. Kidd then explores the role played by 'innocence lost' in the garden refuge, adopting a more complex and 'darker' approach than Cooper's, referencing religious accounts of the Fall and Daoist notions of history as a decline from a state of living in simple harmony with nature. In highlighting the discrepancy between the harmony within the garden and the strife of life outside it, Kidd argues that the contentment we find in the garden refuge is inevitably accompanied by darker 'moods' reflecting a poignant sense of loss. Cooper's 'happy, calm, and unshaken equanimity' is thus scarred by the moral toil of modern life.

In contrast to the perspective of cultivating moral virtues, Alasdair Forbes in Chapter 8, an essay on his own garden, Plaz Metaxu, in Devon, proposes the cultivation of poetic virtues such as bewilderment and melancholy. Unlike Cooper's and Kidd's understanding of nature in classical philosophy as inherently benign, Forbes considers, in line with a lineage of both Asian and Western poets, thinkers and mystics (including Zhuangzi, Ibn 'Arabi, Giacomo Leopardi, Emily Dickinson and the psychologist James Hillman), that nature is at once inviting and bewildering; and in turn, one should learn that the garden is a fragile and tactful intervention in the landscape, that instead of making a binary opposition between light and dark, being and nonbeing, life and death, one should learn to attain a poise between the emergence and the withdrawal of the world. Forbes points out that iconic figures of withdrawal in both Europe and China – citing the examples of the poets Tao Yuanming (365–427) and William Wordsworth – are by no means always as unequivocally serene in their rural retreats as their popular image leads us to expect. Bewilderment and melancholy are as much a part of their worlds as gratitude and contentment. In turning to his own garden, Forbes acknowledges that bewilderment was as instrumental in his reason for withdrawing to make a garden as was gratitude. He describes how bewilderment helps account for the kind of garden Plaz Metaxu is, with particular reference to the Hermes courtyard and the 'interval of emptiness' cradled at the heart of the garden by the valley. He suggests that bewilderment encourages us to 'dwell liminally', or between different horizons of being of the soul. Such liminal openings pierce the security of our garden refuges, in which we must suspend any certainties we may cherish about where we belong.

The garden retreat and mortality

In the final section, two chapters by Yue Zhuang and Michael Charlesworth examine two case studies of the garden retreat centred upon mortality: the Twickenham Villa on the bank of the Thames, in then Middlesex, built by Alexander Pope, an eighteenth-century poet, and the garden at Prospect Cottage on the shingle beach at Dungeness created by Derek Jarman, a modern film-maker. Central to the imaginations of both garden-dwellers was the enigmatic Latin inscription, '*Et in Arcadia Ego*' (even in Arcady am I [death]), and the well-known paintings by Nicolas Poussin (1594–1665), *The Arcadian Shepherds*.[26] An ambivalent image in which the idyllic human existence and its finitude were juxtaposed and contrasted, Poussin's paintings expressed the contemplation of, and preoccupation with, mortality taken up most poignantly by the humanist tradition – a sentiment shared by both garden-dwellers. While the circumstances and their personal beliefs surrounding mortality radically differ (Pope was a devout Christian whereas Jarman was not), there are considerable, interesting parallels. For Pope, the preoccupation with death was affected by his lifelong ill-health and the impact of his belonging to a religious minority; whereas for Jarman, concerns about mortality were brought into sharp focus by the rising incidence of HIV and AIDS within the gay community, to which he had always been actively committed, and, more directly, by his own diagnosis – he died of an AIDS-related illness in 1994, seven years after becoming HIV positive. And despite the 250 years between them, Pope's England was already beginning that process of sidelining death which was to become a hallmark of the modern era. In their different ways of confronting or coming to terms with death, as Zhuang and Charlesworth demonstrate, both Pope and Jarman appealed to esoteric rituals and traditions, as diverse as Christian-Platonic mysticism, Tibetan Buddhism and post-Jungian psychology, which became formative influences in their gardens' spatial organization. The patterns thus developed also served an apotropaic function in warding off the nihilistic side of death and in commemorating the dead.

For Pope, the ritualized itinerary of his garden took the form of a descent through a subterranean passage (the famous grotto) leading to an ascent into an Elysium-like space (the garden proper) which culminated in a reverential processional approach to the obelisk which he had erected in memory of his mother. This itinerary, Zhuang argues, was modelled upon the landscapes of necrotopography (described in classical and Renaissance literature) through which the souls of the dead journeyed and the sacred sites (such as Eleusis) where neophytes were initiated into Arcadian mysteries. Zhuang suggests that, together, the grotto and the garden become a sanctuary which Pope reinvested with esoteric meaning, both revitalizing the Christian concept of death and re-enchanting the garden as a place to dwell, asking us to revise our ideas about the value of mystery in the so-called Age of Enlightenment.

Jarman's use of ritualized structures was first inspired by the Tibetan mandala pattern – circles inside squares inside circles – transposed to his arrangements of stones and plants in the front garden facing the sea. In Tibetan Buddhism and Jung's interpretation of it, with both of which Jarman was familiar, mandalas are 'protective circles', from which one can receive a 'blessing'. Similar protection may also be offered by the seventeenth-century esoteric thinker, Robert Fludd's geomantic pattern, which was adopted in Jarman's design of four circles of planting around four posts assigned to the four orientations of Prospect Cottage. The spatial design of the garden thus assumes the character of an auspicious symbol endowing Jarman with psychological strength. But the real strength, as Charlesworth suggests, was derived from Jarman's memory of 'his dear dead friends', epitomized by the heartfelt understanding that 'each stone has a life to tell', that characterizes his garden. This is where the character of the garden finally lies, in embodying a creativity which not only asks us to reinterpret the surrounding landscape, and the very nature of a garden, but also becomes a courageous and sustaining symbol for those who still survive in the wider (then embattled) gay and queer community.

Notes

1 *Hojoki: Visions of a Torn World*, trans. Yasuhiko Moriguchi and David Jenkins (Berkeley, CA: Stone Bridge Press, 1996). The original title – '*hojoki*' – means nothing more than 'the ten foot square hut' – but the added subtitle 'Visions of a Torn World' in the translation we have used is a faithful reflection of the subject matter of the book and in accord with our topic.

2 See Michael Charlesworth, *The Modern Culture of Reginald Farrer: Landscape, Literature, and Buddhism* (Cambridge: Legenda, 2018).

3 The French publisher Georges Le Rouge emphatically established this conception of landscape gardens with his multi-part *Détail des nouveaux jardins à la mode* (Paris, 1774–89). Several parts of this monumental venture were titled *Jardins Anglo-Chinois.*

4 James Hillman, *The Dream and the Underworld* (New York: HarperCollins, 1979), 67.

5 See in particular ch. 3, 'Psyche', in Hillman's *The Dream and the Underworld.*

6 For Rousseau and reverie, see Michael Charlesworth, *Landscape and Vision in Nineteenth-century Britain and France* (Aldershot: Ashgate, 2008), 81–106. (Despite this book's title, it covers the period 1745–1903).

7 A more detailed account of Forbes's garden of Plaz Metaxu can be found in Alasdair Forbes, *On Psyche's Lawn: The Gardens at Plaz Metaxu* (London: Pimpernel Press, 2020).

8 See Yves Abrioux, *Ian Hamilton Finlay: A Visual Primer* (London: Reaktion Books, 1985), 38.

9 Charlesworth, *Landscape and Vision*, 147.

10 Horace [Quintus Horatius Flaccus], Epistles: Book 1: 18, in *Oxford World's Classics: Horace: Satires and Epistles*, ed. John Davie and Robert Cowan (Oxford: Oxford University Press, 2011), 90. For Pope's use of Horace's motto in his garden, see Yue Zhuang, Chapter 9 in this volume.

11 Abraham Cowley, *Essays*, ed. Henry Morley (London: Cassell & Company, 1893) reprinted at gutenberg.org.

12 D. W. Winnicott, 'Communicating and Not Communicating Leading to a Study of Certain Opposites', in *The Collected Works of D. W. Winnicott*, vol. 6,

1960–1963, ed. Lesley Caldwell and Helen Taylor Robinson (Oxford: Oxford Academic, 2016), 433–46 (439).

13 Diana Balmori and Margaret Morton, *Transitory Gardens, Uprooted Lives* (New Haven and London: Yale University Press, 1993).

14 G. E. Moore, *Principia Ethica* (Amherst, NY: Prometheus Books, 1988 [1903]) 188–9.

15 Wai-yee Li, 'Gardens and Illusions from Late Ming to Early Qing', *Harvard Journal of Asiatic Studies*, 72, no. 2 (2012): 295–336 (304, 324).

16 Ibid., 295.

17 Sophie Volpp, *Worldly Stage: Theatricality in Seventeenth-century China* (Cambridge, MA: Harvard University Asia Center, 2011), 8.

18 Mencius, *Mengzi* (*Mencius*), ch. 'Gaozi I', in the *Chinese Text Project*, ed. Donald Sturgeon. https://ctext.org/mengzi/gaozi-i (accessed 17 March 2024).

19 For an illuminating discussion on this, see Boudewijn Bakker, '*Schilderachtig*: Discussions of a Seventeenth-century Term and Concept', *Simiolus: Netherlands Quarterly for the History of Art*, 23, no. 2/3 (1995): 147–62 (153–4).

20 Friedrich Nietzsche, for example, speaks of this fragmentary experience as a result of linear time in his *Thus Spoke Zarathustra*, trans. Thomas Common, Second Part, Section 20 (New York: Random House, 1917), 15–6. Qing Feng kindly provided this reference.

21 Alberto Pérez-Gómez, *Attunement: Architectural Meaning after the Crisis of Modern Science* (Cambridge, MA: The MIT Press, 2016), 118.

22 Ibid., 183.

23 Gaston Bachelard, *The Poetics of Space,* trans. Maria Jolas, Mark Z. Danielewski and Richard Kearney (New York: Penguin Publishing Group, 2014 [1958]), 34.

24 A sizable scholarship addressed the idea that gardening and other garden practices may help to cultivate virtues and wisdom. See e.g. David E. Cooper, *A Philosophy of Gardens* (Oxford: Oxford University Press, 2006); Robert Pogue Harrison, *Gardens: An Essay on the Human Condition* (Chicago: University of Chicago Press, 2008); Dan O'Brien, ed., *Gardening – Philosophy for Everyone: Cultivating Wisdom* (Oxford: Blackwell, 2010); Yue Zhuang, 'Temple, Huygens and "*sharawadgi*": Tempering the passions to achieve tranquillity', *Studies in the History of Gardens & Designed Landscapes*, 41, no. 4 (2021): 288–308.

25 Martin Heidegger, 'Building, Dwelling, Thinking', in *Poetry, Language, Thought*, trans. Alfred Hofstadter (New York: Harper Colophon, 1975), 148.

26 Poussin painted the same motif twice, first in 1627 and then in 1637–8. For details see Zhuang (Chapter 9) and Charlesworth (Chapter 10) in this volume.

PART I

The garden retreat and theatricality

1

Garden, a refuge and a stage

Death of a garden owner and its moonlit theatricality in seventeenth-century China

Lin Zhu

Introduction: Death of a garden owner

At the end of 1644, Qi Biaojia (1602–45), a competent official and an exemplary literatus, resigned from the court of the Southern Ming Dynasty (1644–62) and returned to his hometown Shanyin, a wealthy county of Shaoxing Prefecture in the Lower Yangzi region. In April of that year, the Ming Dynasty (1368–1644) had collapsed, its capital Beijing was captured by Li Zicheng's (1606–45) peasant rebels, and the emperor committed suicide by hanging himself. Ming loyalists then held the Southern Ming court in Nanjing to continue their loyalist movement in revolt against the Qing reign established by the foreign Manchus, who had overthrown Li and controlled Beijing in June.[1]

There are various perspectives from which to tell the dramatic tragedy of the dynastic succession from Ming to Manchu Qing. The reiterated distinction between 'Chinese' and 'barbarians' signalled the unwillingness of the educated elites to accept the new dynasty – one that, in their view, had been established by a less developed civilization through outrageous conquest. Their resistance was not just a political response to being conquered, but also a reaction to the crisis of culture and tradition that forced them into a life of despair and fear. For Qi Biaojia, a former influential politician whom the new Qing government was eager to bring to heel, the increasing pressure had left him nowhere to hide. One day in July 1645, he had his son warm several cups of wine, and invited relatives and friends to his private garden, the Garden of Mountain Yu. Having told his family that he was going to repair early to his bedchamber, he proceeded to the inner lake, wherein he sat bolt upright and awaited his death. He, thus, became a martyr upon the fall of Ming, at the age of forty-four.

Of all the educated elites who had committed suicide in loyalty to the old Ming Dynasty, Qi Biaojia was among the very few who practised a contemplative and, some might say, elegant suicide during the turmoil.[2] Almost every piece of record, even the shortest and driest ones, does not miss the sense of melancholy and loneliness in his dramatic death, among which the chronicle of Qi Biaojia written by his friend Wang Siren (1574–1646), is the most elaborate one.[3] It poetically depicts in detail the finale of Qi's life staged in his garden, including the dramatic gesture of his midnight suicide in the garden and also his actions before the act.

Stories about Qi Biaojia's obsession with gardens and dramas, his garden life and midnight martyrdom provide a perfect panorama of the literati's world during the Ming–Qing transition, which experienced tremendous change in almost all spheres of life. Besides the economic changes, a socio-cultural revolution, and the beginning of a scientific revolution, cultural practices in this period also reflected the unprecedented and radical transformations in the intellectual world. Among all the new trends and values, dramas and fictional literature and the 'garden mania' were the most impressive ones, which embraced substantial practices and also witnessed the emergence of the discourses.[4] These cultural practices reacted to the chaos and crises of their time by cultivating a deep fascination with illusions and alternative realities, a phenomenon this chapter attempts to illuminate.

This chapter investigates two texts. The first one is Qi's own record of his garden, 'Notes on Mountain Yu', which could be read as reflecting his theatrical imagination of the garden; the second one is Wang Siren's text on Qi's finale in Mountain Yu, when the garden was theatricalized for his farewell. In both circumstances, theatricality serves as the apparatus for constructing the alternative realm of refuge, by the owner and for the owner. Through his elaborate manipulation, we could profoundly comprehend the politics and poetics of his death, where the obsession with theatricality celebrates the human capacity to produce illusory theatrics and freely transcend the opposition between illusions and reality, life and death.

Garden in the master's theatrical imagination

Dramas in the gardens

Before becoming a garden expert, Qi Biaojia gained a prominent reputation in drama criticism in his twenties, having published two anthologies of

FIGURE 1.1 Woodcut print of *Illustration of Mountain Yu*, 1638. Drawn by Chen Guoguang and engraved by Wang Kegong, redrawn by the author based on the original print in *Chorography of Mountain Yu* in the Sonkeikakubunko Collection, Tokyo, Japan. The Chinese characters are names of the scenic spots, but they do not completely correspond to those mentioned in Qi's garden writing, since the painting was finished earlier than the completion of the garden.

Individual scenic spots (listed in order of appearance):

1- Hall of Four Failures 四負堂 (Si Fu Tang)

2- Terrace of Floating Reflection 浮影臺 (Fu Ying Tai)

3- Belvedere of Complete Pleasure* 咸暢閣 (Xian Chang Ge)

4- Pavilion of Great Antiquity 太古亭(Tai Gu Ting)

5- Rock of Cold Cloud 冷雲 (Leng Yun)

6- Gazebo for Friendship with the Rock 友石榭 (You Shi Xie)

7- Retreat for the Return of Clouds 歸雲寄 (Gui Yun Ji)

8- Relinquished to the Gull's Pond 讓鷗池 (Rang Ou Chi)

9- Cherry Tree Grove 櫻桃林 (Ying Tao Lin)

10- Pine Path 松徑 (Song Jing)

11- Water Bright Gallery 水明廊 (Shui Ming Lang)

12- Studio for Reading Book of Changes 讀易居 (Du Yi Ju)

13- Shadows of the Emerald-green Mountains* 蒼山碧影 (Cang Shan Bi Ying)

14- Willow Pathway 柳陌 (Liu Mo)

(Spots marked with * lack explicit evidence for their locations and are therefore speculative based on various sources.)

drama critiques.[5] On his earlier retirement in 1635, Qi stated that he would overcome his addiction to dramas and pick up gardens as his new hobby, therefore started the construction of Garden of Mountain Yu, which lasted for over nine years, until one month before his suicide in 1645.

Mountain Yu sat on a tiny hillock by the river, not far away from Qi's official home in the town. The garden no longer exists, but a sizable amount of literature by Qi and his contemporary remains and enables us to revisit the garden.[6] Qi's own account of his garden, 'Notes on Mountain Yu' (*Yushan zhu*), stands in the centre as the source of subsequent literary production on this garden; abundant details of the garden construction and his casual life are recorded in his diaries.[7] Although having claimed to quit dramas, Qi continued to host lots of drama performances in Mountain Yu; in his diaries, a total of eighty-six plays are mentioned.[8]

The late Ming period witnessed a flourishing of private theatrical performances, after the disappearance of balustrade theatre in the mid-Ming period and before the prevalence of new types of playhouses in the early Qing period.[9] The drama writer and theorist Wang Jide (1542?–1623) listed what he believed made for a suitable environment for the performance of drama, among which 'splendid halls, famous gardens, waterside kiosks, snow-bound pavilions, painted boats, among the flowers, beside the willows' were considered the best.[10] Those were generally garden spots, which provided natural settings and poetic ambience for acting and comfortable places for the audience, thus becoming popular for theatrical performances.

Not surprisingly, Qi's Mountain Yu was one of the most popular places for theatrical performances in his county. Qi and his friends usually enjoyed salon-style drama performances, singing and dancing while having a banquet in the Hall of Four Failures – a large splendid hall facing the inner lake (Figure 1.1-1).[11] The spatial layout of the garden also allowed for a flexible arrangement of outdoor stages and auditoriums. For example, on a clear evening, Qi and his friends stayed at the Terrace of Floating Reflection (Figure 1.1-2), having a casual drink, and then sat in the Belvedere of Complete Pleasure (Figure 1.1-3) to listen to a singer in the Pavilion of Great Antiquity (Figure 1.1-4).[12] The belvedere was located at a higher level on a multi-storeyed structure, which was perfect for the audience, while the pavilion, with a wall-less structure, was appropriate to act as a stage.

Qi's expertise in theatre could suggest that Mountain Yu might have been situated in duality: as a garden and as a stage. In her substantial study of Qi and his garden, Cao Shujuan notices that Qi seems to have projected his sensitivity to theatricals onto his viewing of the garden; in such viewing, the garden was taken as 'an outdoor theatre'.[13] Cao regards Qi's 'Song of the Gentlemen and Ladies Visiting Mountain Yu in Spring' as a play starring young visitors put on at the outdoor theatre, himself being the spectator.[14] Cao interprets this song as five scenes. In the first scene, the young ladies dressed themselves up in the early morning and arrived at the garden; in the second and third scenes, they wandered around different parts of the garden and enjoyed the spring scenery. Then, in the fourth scene, the young gentlemen joined in; they leisurely commented on the beauty of the ladies. Some young gentleman passing the crowd of female visitors dropped his horsewhip purposely, wishing some young lady would notice it; this is a familiar plot of encounter used in love dramas to start a romance, and Qi optimistically anticipated the progression of this story that they would fall in love with each other like a pair of butterflies.

Other literati who contributed to this song collection on the same topic also rendered the potential theatrics and romantics with lively words, even a little jocosity. Impressed by the beautiful flowers, the young ladies picked some and hid them in their sleeves cautiously in order not to be spotted by the gardeners; the path being too narrow, when encountering the males, they attempted to make room for them but accidentally stepped into the grass; the corridors being too long with too many turns, the ladies got lost and puzzled.[15] Their writing closely connected the plots with the unique features of the garden space, as seen in their mention of the narrow paths and curved corridors. Again, such storylines were not uncommon in dramas set in gardens. The theatrical imagination of the garden by Qi and his friends demonstrates another burgeoning trend regarding gardens and theatrics in seventeenth-century China, that is, the flourishing of fiction and dramas in the setting of a garden.[16] In the fictional narratives, the diversiform spaces in a garden trigger the dramatic encountering of characters and the development of plots; meanwhile, the cultural and social aspects of a garden enrich and complicate the meanings that the dramas manage to convey.

Therefore, the entanglement between gardens and dramas in seventeenth-century China should be understood in two senses: on the one hand, a garden is the actual place where a play is performed and enjoyed by the audience; on the other hand, the garden is also the alternative realm to which the performances attempt to transport the audience. If taking the drama as the vehicle connecting the real and the illusory, then the garden presents a duality as it exists on both ends.

The theatrical self and the participatory spectatorship

Deeply engaged with practices on gardens and dramas, it seemed natural that Qi Biaojia had viewed Mountain Yu with his theatrical eyes, as seen in his theatrical imagination of the young ladies visiting the garden. It is noteworthy that such aesthetic sensitivity for theatrics not only manifests in Qi's viewing of human activities but also in that of the non-human garden elements.

Qi had his favourite rock, named Cold Cloud (Figure 1.1-5) after its nebulous shape, and took the rock as his 'little friend' since it 'does not allow himself to become inflamed by the vicissitudes of human affairs and can be said to be my true winter friend'.[17] Qi then built the Gazebo for Friendship with the Rock, sitting opposite the rock, to celebrate their precious and unusual friendship (Figure 1.1-6). In addition, Qi built the Retreat for the Return of Clouds (Figure 1.1-7) for that, '*my* garden has a fine rock named Cold Cloud. Fearing that [the 'clouds'] may "unthinkingly leave the peak", thus failing to stay true to *the Master*'s delight in the mist and clouds, *I* return it to this place for a sojourn.'[18] The retreat did not sit near the rock but close to the entrance of the garden facing the outer waterway, a convenient spot to greet the returning clouds. Hence, centring on the personified rock, Qi created two more sites/characters to participate in the narrative about his friendship with the rock. Their architectural forms and interrelated spatial connections are also shaped by Qi's theatrical imagination of the character relations.

In Qi's noting of the retreat, as quoted above, 'I' and its possessive case 'my' appear in the same sentence as 'the Master' (*zhuren*). Reading it literally, 'the Master' should refer to a third person, a different role from 'I', meaning that 'I' was considering whether 'the Master' could be satisfied. This causes an interesting separation between the master (I) who observes and notes the garden, and the Master whose interaction with the garden is being observed. Duncan Campbell may have also noticed the nuances between this term and the more commonly used 'I', therefore capitalizes the first letter of the phrase 'the Master of the Garden' in his English translation of the 'Notes'.[19]

This special term, 'the Master', appears in eleven out of the forty-nine notes. In those scenes, the Master emerges and interacts with the personified garden elements, in a narrative focusing on the emotional connections between the characters. In 'Relinquished to the Gull's Pond', for example, Qi describes how the setting sun illuminates the surface of the pond in an incredible scene in which 'the Master calls joy indeed', while the gulls 'do not see the difference' (Figure 1.1-8). Therefore, the Master of the Garden dares not lay claim to the pond for himself but rather relinquishes it to the gulls. In doing so, however, he fears that the gulls will suspect his motives and may not accept his gift.[20] Another example could be found in the section 'Cherry Tree Grove'. Qi describes how when the blossoms fall as snow and the red fruit hangs like stars from their branches, it is like 'catching a glimpse of a great beauty from behind a screen, her red lips only half revealed' (Figure 1.1-9). However, the Master of the Garden happens at that time to discuss abstruse matters with Laymen from Mount Culai (referring to the pine trees on the nearby the Pine Path, Figure 1.1-10), therefore does not need the beauty and her romantic songs. Qi plays with different levels of affinity with the sites in his garden.

The poetic quality of these texts dissimulates the subtle change of the narrative subject. 'The Master' becomes part of the spectacle when his emotional connection with the rock, the gulls, the cherry trees and the pine trees is spectated by the master. Its emergence marks a moment when the 'theatrical self' is separated from Qi Biaojia and allows him to turn his gaze upon his alter ego. The capacity for simultaneous impassioned participating and dispassionate viewing makes him an omniscient observer, whose spectatorship is both participatory and panoramic. Thus, he could embrace sensual involvement and detached observation, although playful, at the same time.

Sophie Volpp takes 'participatory spectatorship' as an important feature typifying the theatricality in seventeenth-century China, since 'the perspectival challenge that asks the spectator to hold illusion and disillusion in tension simultaneously, however, could be considered quintessentially theatrical'.[21] She further elaborates that such spectatorship is premised on the mobile vision and provisional theatrical space, which, in Qi Biaojia's case, were guaranteed by his free manipulation of spectatorship as the master of the garden.

The participatory spectatorship in Qi's 'Notes' seems to have been spotted by his friend Zhang Dai (1597–1684), an essayist, historian, and connoisseur in dramas labelled as 'the most refined spectator' by Volpp.[22] Zhang wrote a postscript to 'Notes on Mountain Yu' wherein he commented admiringly that 'when the master notes his garden . . . he is able to catch the subtlety that is known exclusively to himself, not to others'.[23] Zhang then explained this with an old story. The poet and artist Su Shi (1037–1101) visited the Buddhist frescos in a temple late at night, when his lamplight almost went out. Looking at the frescos closely, moving inch by inch with 'the weak and flickering lamp', Su sensed that 'the Buddha in the painting seemed to be moving'. Zhang noted that it was not the exquisiteness of the painting itself but Su Shi's specific way of viewing under the dying lamplight that connected him with the painting. It created a dynamic spectatorship – mobile and mutable, with Su himself participating in the spectacle as the moving lighting. By comparing Qi's viewing of his garden to Su's night tour of the frescos, Zhang Dai intends to highlight the exclusiveness of participatory spectatorship; their exclusive comprehension of the observed objects does not rely on their privileged knowledge of the object but the secret manipulation that could be 'exclusively known to himself'.

Late Ming moonlit theatricality

In Su Shi's nocturnal view of the frescos, he was not as proactive as Qi in pursuing participation in spectacles, but the nocturnal obscurity sent him there. It was common in late Ming dynasty that private theatrical performances in private venues were held at night. In Mountain Yu, except for those specially arranged for Qi's aged mother, most performances took place at night and even midnight, and 'ended after we had the ultimate pleasure', as Qi recorded. The well-known playwright, novelist and publisher, Li Yu (1610–70), elaborated on his preference for nocturnal performances for two reasons:

> [Firstly,] the performances do not inform realities. Their exquisiteness lies in obscurity. If dramas are played during the daytime, it will be too distinct for the actors to create the illusion with their acting skills. Only half of their voices and facial expressions could be properly perceived since the acting is largely distracted. Secondly, everyone, whether rich or poor, noble or low, has work to do in the daytime. When the play starts at dusk, both the hosts and the guests are at ease and free from any anxiety of failing things. This is why the ancients enjoyed night tours with candles.[24]

The first reason considers the lighting environment on-stage, and the second refers to that condition offstage. For Li Yu, obscurity is necessary for both, especially in private venues, where the demarcation between the stage and the audience can barely be established due to the limited space and its provisionality. The natural and social obscurity offered by the nocturnal ambience, helped the actors to act convincingly and, more importantly, distanced the audience from the disturbances of mundane affairs, both in a perceptual and psychological sense, so that they could be totally immersed in the illusory world. When performances had to be played in the daytime, hosts of the private venues, such as Ruan Dacheng (1587–1646), a dramatist, poet and official, would set up layers of curtains enclosing the venue and lit decorated torches inside.[25] The curtains blocked external distractions and constructed an isolated realm of obscurity, allowing the scene to function in a similar way to the night.

Li Yu's preference for nocturnal performances considers the specificity of private venues where temporal stages were set for performances, and the nocturnal obscurity that empowers the departure from the daily realm of distinction, reality and work. In this sense, the non-daytime and non-daily environment matter significantly for the delivery of things that 'do not inform realities', which include but are not limited to drama performances.

Qi's diaries record a sizable number of nocturnal events in Mountain Yu when he invited friends to night parties and enjoyed theatrical performances, after which sometimes they sat on an open terrace to watch the fireworks.[26] Lamps and lanterns were commonly used in nocturnal activities in the late Ming, especially in Buddhist services. Hanging lanterns in the wooded mountain and arranging lamps on the water surface or along the bank were Qi's favourite activities and at times he connected these scenes to Buddhist realms. In his note on the Terrace of Floating Reflection, he compares the open terrace illuminated by moonlight and surrounded by lotuses to the Buddha's seat in the shape of a lotus flower (Figure 1.1-2). When noting Water Bright Gallery, the first scenic spot entering Mountain Yu, Qi describes how the water reflects the moonlight onto the visitors in the gallery, so that 'both the Master and guests look as if they have travelled from the Kingdom of Glazed Glass; their beards and eyebrows seem to drip and their clothes are drenched' (Figure 1.1-11).[27] The allusion to the Kingdom of Glazed

Glass not only helps to render the scene but also implies the delight to be transported to the Buddhist realm of the pure land. Here, we catch another moment where the master himself sees the alienated Master.

John Dardess also observes the increasing interest in nocturnal enjoyment from his study of the landscape in Tai-ho County, Jiangxi Province. He points out the crucial change in landscape appreciation in late Ming:

> From the early Ming discovery of beauty and value in the real world of nature and life on the land, one ends in the late Ming and early Qing determination to ignore or avoid that world, with the creation of toy fantasies in artificial lakes and landscaped gardens, and a preference for moonlit surrealisms to the daylight world of settlement, work, and subsistence.[28]

This tendency seems to echo Li Yu's preferred environment for theatrical performances: non-daytime, non-daily, and ready for displacement from reality. In this sense, Qi Biaojia's garden life, as recorded in his diaries and garden writing, offers a perfect footnote to this change. Compared with his daytime in Mountain Yu, when he was occupied with daily affairs, such as supervising the garden construction, completing the accounts, cultivating the vegetation, etc., the night-time activities offered opportunities for nocturnal enjoyments, artificial engagements and sentimental reactions. The intention to manipulate and celebrate such moments speaks to a burgeoning trend in the late Ming period: it was less a case of internal 'feelings' (*qing*) meeting with the external 'landscape' (*jing*), which had been the essential basis of Chinese landscape aesthetics, but rather, the external world was itself displayed to stimulate certain senses and feelings.[29] The popularity of artificial intervention in the natural landscape and designated ways to appreciate it suggests that the imagination of landscape surpassed the natural beauty itself; as Dardess observes, 'nature found itself modified and to a significant degree displaced by artifice'.[30]

Summing up his study, Dardess attributes the changing perceptions of the landscape to the changing social pressures of the time. After the more optimistic sixteenth century with 'local security and decent opportunity', the literati's world was replaced by 'a more limited, more problematical, and more distasteful one' in the seventeenth century.[31] Qi Biaojia would probably agree with Dardess's inference; in his letter to a friend in 1637, he admitted that the most severe problem of the nation was 'neither the foreign invasion [of the Manchus] nor the domestic bandits [in the peasants uprisings]', but the existing problems inside the Ming court and society.[32] Qi reveals his profound understanding of the transience and temporality of the human world in his noting of Studio for Reading the *Book of Changes*, a place where he studies the *Book of Changes* to get rid of his weariness of mundane affairs (Figure 1.1–12). Qi notes that when along the bank the lamps are lit in the night and music is played, the Master feels the scene to have been

Heaven-sent, on which he immediately reflects that 'Even Heaven and Earth cannot determine its fortune of completion or devastation'.[33] The manipulated illusions in gardens not only transport the master from reality to illusions of refuges; meanwhile, they become the basis for him to contemplate the fugitive existence of things and the possible loss.

For this reason, gardens are extensively involved in the production and manipulation of illusory refuges in seventeenth-century fiction, dramas and other miscellaneous writing, as Li Wai-yee notices, 'the aesthetics of illusion in gardens . . . overlapped with dreams, memories, or fantasies wherein a lost world could be reclaimed or an escape from an alienating reality invented in the post-conquest world'.[34] In this way, the garden makes a significant refuge for the post-conquest life of the Ming literati; and in Qi Biaojia's case, it even makes a stage for his death.

Finale of the garden owner: Theatricals between life and death

Now it is time that we fix our sights back on Qi Biaojia's midnight suicide. The elaborated record by Wang Siren portrayed a manipulated finale, displayed at the provisional stage of a garden and ended at dawn when the sun came out. The finale started when the night fell: Qi walked to an outdoor corridor after the banquet with his family and friends, and it was dim (Figure 1.1-13):

> Looking at the southern mountains, he smiled and said, 'Mountains and rivers are all illusions, as well as human beings; never do mountains and rivers change, while quickly another lifetime passed.' Entering the room, Qi pointed to the zither scores and said, 'These are all traces of the past!' . . . He demonstrated the gesture of hanging himself with his hands, . . . sitting with his eyes closed for a good while, and then opened his eyes and said, 'Is it really difficult to die? If it is just like this, I shall not hesitate.'[35]

The nocturnal dimness brought Qi's world into illusions. When he compared the existence of a human being with the eternal mountains and rivers, he was acknowledging the transience of life; when he saw the zither scores as 'traces of the past', he was pretending to be a posthumous soul witnessing his own death and his remains; when he fictively hung himself or sat with eyes closed, he was attempting to feel what it was like to die to persuade himself that dying was not as difficult as people had imagined. Those performative actions reveal a sentimental theatrical role that exists alongside the surface role of his sensitivity to the obligated loyalty, thus providing another perspective from which to understand the dilemma of a martyr in the dramatic situation of the Ming-Qing transition, and his decision.

FIGURE 1.2 The view of the southern mountains from the original site of Mountain Yu in Shaoxing (now in Zhejiang Province). In Qi Biaojia's time, the profile of the mountains should be more distinct, without the intrusion of dense urban construction in the midground. Source: Author.

Late Ming literati articulated the idea that the world is a theatre and that life itself is a play, because theatre could be a means of interpreting almost any aspect of life.[36] Inspired by the evolving Buddhist thoughts on the theatrical nature of the human world in this period, the literati came to realize that this association was more than literary rhetoric but was truly instructive in the practice of transcending suffering and death. Those who truly understand the theatrical nature of life can achieve Buddhist enlightenment, as well as freely transcend the limits of life and death; as the Buddhist master Juelang Daosheng (1592–1659) summarized, '[the enlightened] come out of death and enter life, playing in both the mundane and sacred realms'.[37] With the theatrical gestures and speech before his suicide, Qi repetitiously immersed himself in scenarios where he was dying or already dead. The 'theatrical self' emerged in this scene rehearsing the forthcoming death and allowed his strategy of making everything equally illusory to manifest so as to cross the boundary between life and death repeatedly, and consequently, obfuscate their demarcation.

Qi's significant counterpart from Shaoxing, Ni Yuanlu (1594–1644), successful as a literatus and a politician, was also addicted to enormous garden construction. Ni likewise displayed some theatricality in his finale in his garden, Belvedere Dressed in Clouds, a three-storey belvedere facing Mountain Qinwang, one of the highest mountains in Shaoxing, with thousands of bamboos planted on the third floor.[38] Before Ming's fall, Ni Yuanlu's mentor Huang Daozhou (1585–1646) visited the belvedere, finding that Ni had 'decorated it with brocaded draperies and hung lamps to illuminate all places' to greet him. Huang was angry because no banquets or entertainment would be appropriate when the nation had been going through such difficulties, while Ni said laughing, 'This is to bid farewell to you'. Soon, Ni recruited troops with his family fortunes and headed north to Beijing. After the emperor hanged himself, Ni also hanged himself in his house.[39] Another example of a pre-death carnival could be that of the poet-official Xia Yunyi (1596–1645), who had been recruiting troops to fight against the Qing until his martyrdom. Foreseeing the failure, Xia spent all his savings preparing a sumptuous feast with opulent food and wine for guests from a wide range of social statuses. When the guests were drinking, laughing, singing and playing games, Xia stood up, pretending to go to the bathroom, while instead going straight to drown himself in the nearby river.[40]

Through the dramatic acts before their principled deaths, Qi, Ni and Xia prepared themselves for the departure from life. Such preparation took place in the night, with the deliberate construction of non-daily scenarios, bringing the moment into a sentimental realm of immersive experiences, where they themselves were clear about the underlying intention of the manipulation. In the profound understanding of transience and ephemerality by the late-Ming literati, the garden could be seen as a stage, the world as the extension of that stage, and life as a play. Although committed in solitude, their suicides had a historical audience. In the political context of the dynastic transition, it was important to note the theatricality in their martyrdom that was meaningful to external viewers – Qi's body was spotted the next day at dawn:

> The east began to brighten and revealed inches of his square kerchief at the waterside around the stone steps in front of the Belvedere of Plums (close to the Pillow Pathway) (Figure 1.1-14). When they hurried over, they found Qi sitting just under the water's surface in a composed posture, with his hands down and his feet together. His forehead was submerged, while his hat and hair remained orderly. His face appeared smiling.[41]

The way Qi Biaojia killed himself – sitting and smiling, without suffering and struggling – is by no means a rash decision but something that was deliberately choreographed. His decent and orderly death seems to have challenged physiological facts, but made perfect moral sense for a political death, by exaggerating his determination and calmness in confronting death. Contemporary scholars have noticed the presence of theatricality and the existence of external spectators in the cultural practices of the late Ming period. In Richard Vinograd's reading of Chen Hongshou's (1598–1652) portraiture, he demonstrates the 'theatricality' of those paintings, including the 'masklike exaggeration' and 'conscious posturing' of the figures, which indicates their awareness of the external viewers, and the analogy of the pictorial space to the stage.[42] The theatricality is related to 'an enhanced consciousness of the multiplicity of roles available to the self' in seventeenth-century China, as Volpp observes in the contemporary study across the disciplines on late Ming art; the scholars have remarked on a new sense of an 'I' and are aware that their multiple roles are spectacles visible to others, which is evident in their 'fascination with self-dramatization, masking, fabrication, and dissimulation'.[43]

Their investigation implies another lens to examine the theatricality of Qi Biaojia's death. My earlier analysis – of the theatrical spectatorship of his garden and theatrical actions prior to the suicide – mainly examines the issue of theatricality from the perspective of *self-expression*; the theatricality lies in his manipulation of illusions and his play with the boundary of illusions, which basically manifests in the private realm. However, it is also important to note the external gaze of the manipulated theatricality, when the social and historical spectators spectated his actions, inner thoughts and feelings, which would otherwise remain unknown. Viewed from this point, we could not miss the mention of the partially visible kerchief above the water, which guided his family to finally find his body. The kerchief-hat was the most typical part in the Han hairstyle of the male and had become a symbol of loyalty since the 'Queue Order' was forcefully enacted by the Qing government before Qi committed suicide. It was at this moment that the greatest resistance activities broke out in the Lower Yangzi region.[44] The theatricality then comes to the public realm and *represents* the juxtaposition of the poetics of a literatus' sentimental farewell and the politics of a martyr's death.

Philosophy of the post-traumatic living

Qi Biaojia's garden was named after the small hill where the garden was located. *Yu* originally means to lodge, to reside. Decoding *yu* is an important theme in the writing of Mountain Yu among all the authors. Recounting his visit to Mountain Yu, Wang Siren wrote: 'The stars lodge in the sky, the mountains and rivers lodge on the earth; the sky and the earth lodge in vitality; for human beings, their souls lodge in their mind-hearts, and their mind-hearts lodge in their body. They lodge wherever they go.'[45] Wang Shimei also echoed this theme, stating that, 'on the one hand, a long stay could not be called lodging; on the other hand, even a long stay is still lodging'.[46] Qi and his friends seem to have embraced the temporality of inevitable changes and hence the 'lodging' of everything; regardless of how substantially existent something is, or how long it may stay, they came to view its essential condition to be one of lodging.

The concept of *yu* embodies the loose and unstable dependency between things and the world they live in. As the owner, Qi had accepted his temporal ownership of his garden and prepared himself for the foreseeable loss. He even negated his identity as the master of the garden by stating that 'whoever lodges is the guest and by no means the master, so I am also a guest as you are'.[47] The dialectical opposition between the master and the guest reminds us of the juxtaposition of 'I' and 'the Master of the Garden' in Qi's 'Notes'; when the immediate alienation of the 'theatrical self' posited him in the bivalent role of spectacle–spectator, he felt himself to be just one guest of the garden among many. Hence, the garden serves as a medium for Qi Biaojia to explore the multiple dualities in the world – dreams and waking, reality and illusions, the master and guests, the spectator and spectacles – all of which manifest in shaping the relationship between an individual and the living world, while the garden presents the unstable boundaries between these opposing elements, where one could be easily transited in between.

John Minford notices the magic of the transition in private gardens after examining sizable fiction which takes place in gardens in the early Qing:

> The garden in its various forms is again a halfway house. It is here that characters encounter the strange, the supernatural, the erotic. Within the house normality prevails, but once out in the open space, down a walkway or an overgrown temple cloister, and the transition can all too easily be made.[48]

In this sense, a place to lodge equals 'a halfway house'. It provides physical and mental protection from the calamities of life; however, when spectated as a stage, the quotidian familiarity of the house gives way to the immediate theatricality of the garden, allowing the immediate 'theatrical self' to emerge and perform, thus suspending the garden in illusions.

In the end, there emerges a perfect socket for all the dualities discussed in this chapter – in many senses, Mountain Yu is a halfway house. If taking the stage as one of the various antitheses of the house, the garden stays midway between publicity and privacy, reality and illusions, ordinary and spectacles, daytime and moonlight, and so on. The theatricalized garden provides a stage for self-expression and representation, which, when examined in the context of political upheavals, reflects a philosophy of living in the post-conquest world. For Qi Biaojia, the ideal of a solid home had been crashed, and the theatricalized garden became part of his fugitive existence, even a symbol.

Notes

1 For a comprehensive overview of the Ming-Qing transition, see Frederic E. Wakeman, *The Great Enterprise: The Manchu Reconstruction of Imperial Order in Seventeenth-century China* (Berkeley: University of California Press, 1985), vol. 1, chs 1–8.

2 An official record in early Qing, 'Imperial-ordered Record of the Martyrs of the Preceding Dynasty', recorded that more than 3,883 Ming loyalists died for the nation after Ming's fall in 1644. Most martyrs were civil service officials and influential literati. Their death should not be understood as merely personal choices, but also following the indoctrination of Confucian loyalty that a

decent man cannot serve two dynasties. For the dramatic choices of literati after the Ming-Qing transition, see Frederic E. Wakeman, 'Romantics, stoics, and martyrs in seventeenth-century China', *The Journal of Asian Studies*, 43, no. 4 (1984): 631–65.

3 Wang Siren was Qi Biaojia's fellow Shaoxing native who starved himself to death after the downfall of his hometown. As a close friend of Qi, he had visited the Garden of Mountain Yu many times and contributed to the literature of the garden.

4 Handlin Smith uses the word 'mania' to describe the 'real momentum, both in the quantity of gardens and in the amount of documentation they generated' in the late Ming, especially in those prosperous areas. See Joanna F. Handlin Smith, 'Gardens in Ch'i Piao-Chia's Social World: Wealth and Values in Late-Ming Kiangnan', *The Journal of Asian Studies*, 51, no. 1 (1992): 57–8. In relation to the flourishing of gardens and dramas, there accumulated rich traditions of lore and specialized manuals or catalogues for connoisseurship. For example, Ji Cheng (1582–1642) published *The Craft of Gardens* (*Yuan ye*) in 1631, which is now considered the first and definitive work on garden design in China; and in 1624 Wang Jide (1542?–1623) published *Principles of Lyric Drama* (*Qu lü*), a monumental theoretical book of drama aesthetics.

5 Qi's two anthologies of drama critiques were important in that they covered an unprecedented number of dramas; some of them had never been mentioned in others' works. Qi was also a playwright, although none of his works remains.

6 Qi Biaojia compiled *Yu shan zhi* (Chorography of Mountain Yu) to include the exhaustive collection of writings on his garden in various forms of literature. Over 200 literati contributed almost 1,000 pieces of works in total, with more commentaries affiliated.

7 Texts from Qi's 'Notes' in this chapter refer to the modern reproduction in Zhao Haiyan, *Yushanzhu yanjiu* [A Study on 'Notes on Mountain Yu'] (Hefei: Anhui jiaoyu chubanshe, 2016), 336–92; the English translation is modified from Duncan Campbell, '"Footnotes to Allegory Mountain": Introduction and Translation', *Studies in the History of Gardens & Designed Landscapes*, 19, nos. 3–4 (1999): 243–71. Qi kept diaries for fifteen years from 1631 to 1645, which were compiled and reprinted in 2017, see Qi Biaojia. *Qi Biaojia riji* [Diaries of Qi Biaojia]. Hangzhou: Zhejiang guji chubanshe, 2016.

8 Zhao Suwen, *Qi Biaojia yanjiu* [Studies on Qi Biaojia] (Beijing: Zhongguo shehui kexue chubanshe, 2011), 138–206. Zhao has a comprehensive study of the plays that Qi Biaojia watched and mentioned in his diary.

9 The balustrade theatre refers to the commercial theatre built amidst marketplaces, brothel districts, and other urban environments with active commerce and heavy traffic. For the development of private theatres in the late Ming, see He Yuming, *Productive space: Performance texts in the late Ming* (PhD Dissertation, University of California, 2003), 29–49.

10 Wang Jide, *Qu lü* [Principles of Lyric Drama] (Changsha: Hunan renmin chubanshe, 1983), 274.

11 Taking the year 1637 as an example, after the hall was accomplished, such banquets were recorded in Qi, *Qi Biaojia riji*, 264, 276, 293.

12 Ibid., 591. The Belvedere of Complete Pleasure was built after the painting was finished, therefore was not in the *Illustration of Mountain Yu*. Instead, I marked another belvedere (Figure 1.1-3) to show the multi-storeyed structure.

13 Cao Shujuan, *Liubian zhong de shuxie: Qi Biaojia yu yushan yuanlin lunshu* [Writing in Turmoil: Qi Biaojia and His Discourse on Mountain Yu] (Taipei: Liren shuju, 2006), 281–91.

14 The collection of songs is compiled in *Yushan youyin* (Verses of Visiting Mountain Yu) in the *Chorography*. The Song by Qi Biaojia is quoted from the modern reprint in Qi Biaojia, *Qi Biaojia ji* [Complete Works of Qi Biaojia] (Beijing: Zhonghua shuju, 1960), vol. 9, 222–3.

15 The three extracts mentioned here are from the songs by Zhang Dai, Zhang Hong and Wang Siren, quoted from the secondary resource; see Cao, *Liubian zhong de shuxie*, 288–9.

16 Contemporary scholarship on gardens in Chinese dramas and fiction in the seventeenth-century is substantial. Gardens are viewed not only as physical spaces where the story takes place but also as cultural and social spaces imprinted by the transformation and crisis of the society. See Mary Scott, 'The Image of the Garden in Jin Ping Mei and Hongloumeng', *CLEAR*, 8 (1986): 83–94; Xiao Chi, *The Chinese Garden as Lyric Enclave: A Generic Study of the Story of the Stone*, no. 93 (University of Michigan Centre for Chinese, 2001); Jin Lei. 'Erotic Enclaves and Contested Beds: Gardens in Pu Songling's Chuanqi Tale', *ASIANetwork Exchange: A Journal for Asian Studies in the Liberal Arts*, 22, no. 1 (2015): 23–37.

17 Qi, 'Yushan zhu', 349.

18 Ibid., 347–8. The site name alludes to the couplets 'the clouds, unthinking, leave the peaks; the birds, tired of flying, know it is time to return' in Tao Qian's 'The Prose-poem of Returning Home', the famous poem advocating retiring from public life and returning to rustic simplicity.

19 Campbell, 'Introduction and translation', 243–71. Extracts with 'the Master of the Garden' can be found on 248, 251, 252, 256, 259 and 262.

20 Qi, 'Yushan zhu', 341.

21 In Volpp's study, she uses the term 'participatory spectatorship' in two senses: the theatrical and the social; here my reference to her study mainly focuses on the theatrical aspect. See Sophie Volpp, *Worldly Stage: Theatricality in Seventeenth-century China* (Cambridge: Harvard University Asia Center, 2011), 8, 11, 32–40.

22 Ibid, 8.

23 Zhang Dai, 'Ba Yushan zhu', in *Langhuan wenji* [Literary Collection in Langhuan] (Changsha: Yuelu shushe, 1985), 210–11.

24 Li Yu, *Li Yu quanji* [Complete Works of Li Yu] (Hangzhou: Zhejiang guji chubanshe, 1991), 71.

25 Jiao Xun, *Ju shuo* [Discourse on Dramas] (Shanghai: Gudian wenxue chubanshe, 1957): 128. Ruan Dacheng was also a significant figure in the garden culture of the late Ming period. He owned multiple gardens, supported the garden designer and theorist Ji Cheng (b. 1582) and published Ji Cheng's

book *Yuan ye* (*The Craft of Gardens*) in 1631, which is now considered the definitive work on Chinese garden design.

26 Qi, *Qi Biaojia riji*, 288, 293.

27 Qi, 'Yushan zhu', 338. The Kingdom of Glazed Glass refers to the pure land of the Medicine Buddha, the Vaidūryanirbhāsā.

28 John W. Dardess, 'A Ming Landscape: Settlement, Land Use, Labor, and Estheticism in T'ai-Ho County, Kiangsi', *Harvard Journal of Asiatic Studies*, 49, no. 2 (1989): 363.

29 For a chronicle study on this aesthetic tradition and its specific influence on landscapes and gardens, see Zou Hui, '*Jing* (景): A Phenomenological Reflection on Chinese Landscape and *Qing* (情)', *Journal of Chinese Philosophy*, 35.2 (2008): 353–68.

30 Dardess, 'A Ming Landscape', 357.

31 Ibid., 363.

32 Zhao, *Yushanzhu yanjiu*, 275.

33 Qi, 'Yushan zhu', 339. The *Book of Changes* is an ancient Chinese divination text and among the oldest of the Chinese classics – an influential text inspiring religion, philosophy, literature and art. The 'Notes' reflected Qi's specific interest in this book from various perspectives; see Cao, *Liubian zhong de shuxie*, 157–9.

34 Li Wai-yee, 'Gardens and Illusions from Late Ming to Early Qing', *Harvard Journal of Asiatic Studies*, 72, no. 2 (2012): 334.

35 Wang Siren's chronicle for Qi Biaojia has been collected and annotated in the modern reprint of Qi Biaojia's diaries. See Wang, 'Qi Zhongmingong nianpu' [Chronicle of Duke Qi Zhongmin], in *Qi Biaojia riji* (Hangzhou: Zhejiang guji chubanshe, 2016), 845–934. This quotation is on 891.

36 The association of drama and human life prevailed in seventeenth-century China, not only in drama and fiction but also in other literary works. See Zhao Yingzhi, *Realm of Shadows and Dreams: Theatrical and Fictional Lyricism in Early Qing Literature* (Harvard University, 2014), 116–20. Zhao has several examples in early-Qing literature to develop her argument on the influence of theatrical culture on the metaphor of life and its interpretation.

37 Juelang Daosheng, *Tianjie juelang chanshi quanlu* [Complete Works of Master Juelang], vol. 25, 24–5. Buddhist monks using the metaphor of life as a play to enlighten listeners has appeared since the Song Dynasty and was developed in late Ming, under the reinvention of Chan Buddhism and the prosperity of dramas. See Liao Zhaoheng, 'Chanmen shuoxi: yige fojiao wenhuashi guandian de changshi' [Discoursing dramas with Buddhist thoughts: an attempt from the perspective of Buddhism cultural history], *Hanxue yanjiu* [Chinese Studies], 17 (1999): 277–98. Qi could have been influenced by Juelang's philosophical thinking on dramas, since according to his diaries, he had read *Quotations of Master Juelang* in 1640 and had personal correspondences with Juelang in 1645.

38 Ni Yuanlu resigned from the Ming court in 1635, the same year as Qi Biaojia, whereafter they had quite a lot of gatherings in their gardens and other casual

places in Shaoxing. His Belvedere Dressed in Clouds was accounted by several contemporary literati, including Qi Biaojia. The description here refers to Huang Zongxi. *Huang Zongxi quanji diyice* [Complete Works of Huang Zongxi, vol. 1] (Hangzhou: Zhejiang guji chubanshe, 1985), 343.

39 Zhu Yizun, *Jingzhiju shihua* [Remarks on Poetry from Jingzhiju] (Hangzhou: Fuli shanfang, 1819), vol. 12, 14b–15a. Huang Daozhou is a calligrapher, scholar and official who was killed by the Manchu army for his refusal to surrender.

40 Chen Zhenhui, *Shanyang lu* [Records of Shanyang] (Reprint, Taipei: Xin wenfeng chuban gongsi, 1988), 707. Xia was mentioned a lot in Qi Biaojia's diaries but their communication was mostly through letters sharing opinions on governance. Xia had not visited Mountain Yu.

41 Wang, 'Nianpu', 892. The Belvedere of Plums is not seen in Figure 1 nor Qi's 'Notes'. Details vary in the records on Qi Biaojia's death and different spots are mentioned. According to the author's investigation on these versions, the spot should be around Willow Pathway, as marked in Figure 1-14.

42 Richard Vinograd, *Boundaries of the Self: Chinese Portraits, 1600–1900* (Cambridge: Cambridge University Press, 1992), 31–3. Chen Hongshou excelled in painting and calligraphy and was best known for his peculiar portraiture works. Liu Shi-Yee further articulates the theatricality of Chen's figure painting with more of Chen's works in Liu, 'The World's a Stage: The Theatricality of Chen Hongshou's Figure Painting', *Ars Orientalis*, 35 (2008): 155–91.

43 Volpp, *Worldly Stage*, 17.

44 Two weeks before Qi's suicide, the regent Dorgon issued an edict ordering all men in Ming to shave their foreheads and braid the rest of their hair into a queue identical to those worn by the Manchus. Ming people resisted the order and the Qing struck back with deadly force, massacring all who refused to obey.

45 Wang Siren, 'You Yushan ji' (Account of visiting Mountain Yu), in *Chorography of Mountain Yu.* Quoted from the secondary source; see Zhao, *Yushan zhu yanjiu*, 296.

46 Wang Shimei's annotation on Zhang Dai's 'Yushan ming' (Inscriptions on Mountain Yu); quoted from the secondary source. Ibid.

47 Chen Qiyuan, 'Yushan wen' (Inquiring about Mountain Yu), in *Chorography of Mountain Yu.* Ibid.

48 John Minford, 'The Chinese garden: Death of a symbol', *Studies in the History of Gardens & Designed Landscapes*, 18, no. 3 (1998): 260.

2

Daguerre's 'natural diorama'

Landscape design and picturesque retreat in nineteenth-century Bry-sur-Marne

Kasie Alt

He conceived of it as a natural diorama . . . It was a charming spot that showed with what wise combinations Daguerre knew how to simulate nature.[1]

The 'natural diorama' Adrien Mentienne (1841–1927) described above in *Découverte de la Photographie* (1892) refers to features designed by Louis-Jacques-Mandé Daguerre (1787–1851) in the landscape of the château in Bry-sur-Marne. A historian and archaeologist of the region, Mentienne was born in and eventually became mayor of Bry-sur-Marne. His text, which is primarily dedicated to establishing Daguerre's primacy in the invention of photography, has several moments where he discusses Daguerre's involvement in the life of the town. While these vignettes establish and reinforce connections between the well-known man and this Paris suburb, they also reveal the artist's significant role in the life of the town. While Daguerre's work in the château landscape creates a retreat for its owner, it also participates more publicly in the life of the village by employing local workers during and after the French Revolution of 1848.[2] The château landscape at Bry-sur-Marne has received little scholarly attention to this point due to a paucity of archival evidence and that the landscape no longer exists as such. Yet this collaborative effort between Daguerre and other notable citizens of Bry-sur-Marne created a distinct space that leveraged the spectacular nature of the Diorama[3] and the aesthetics of the picturesque as an antidote to the social and political upheavals that resulted in the end of the Orléanist Restoration.

Located just outside Paris, the château estate at Bry-sur-Marne created a picturesque evocation of nature framed by the illusions of the Diorama. Intended to evoke the informality of nature, the landscape was further elevated through the inclusion of several decorative *fabriques*, garden follies, including a grotto near the northern boundary of the estate. The use of such *fabriques* and simulations of nature are well-noted features in landscape designs before and after 1848.[4] In pre-revolutionary France, some of the most celebrated examples include Ermenonville, Méréville and Jardin Monceau. After 1848, the works of Jean-Charles Adolphe Alphand (1817–91), such as the Buttes Chaumont and the Bois de Vincennes, employ similar picturesque aesthetics as framed by decades of landscape imagery, tourism, and entertainment.[5] The château landscape at Bry-sur-Marne indicates a watershed moment between the pre-revolutionary landscapes and the later works of the Second Empire. While the landscape's impact was mitigated due to its disappearance just ten years after Daguerre's involvement, it offers a unique window into the use of landscape aesthetics in response to the political events of the time. As a direct response to social and economic difficulties at the end of the July Monarchy, Daguerre's work in Bry-sur-Marne's château landscape manifests an aesthetic rooted in ideological expectations of nature and illusion formed by the Diorama spectacle.

Daguerre is best known today as the eponymous co-inventor of one of the earliest photographic processes, the Daguerreotype.[6] Scholarly narratives follow his artistic training and entrepreneurial ambitions in Paris, and tend to wrap up neatly by noting that he retired to his *maison de campagne* (country house) in Bry-sur-Marne by the 1840s.[7] Yet records indicate that he did more than retire there at the end of his life. During his time in Bry-sur-Marne, Daguerre involved himself in the affairs of the town, befriending the then-mayor Armand-Louis Mentienne (1798–1857) and becoming a *conseiller*, a town councillor, in 1843.[8] His photographic and artistic experiments were further supported by the local *châtelaine*, the owner of the local château and former seat of the *seigneurie* of Bry-sur-Marne, Mademoiselle Geneviève de Rigny (1775–1857).[9] By the time Daguerre took up residence in Bry-sur-Marne, Mlle de Rigny owned the château and its lands outright as an inheritance from her uncle, Joseph Dominique, also known as Baron Louis (1755–1837).[10]

Mentienne's description of the château landscape as a 'natural diorama' places it in the context of Daguerre's other works.[11] Before his photographic experiments earned him the *Légion d'honneur*, Daguerre's primary claim to fame was the Diorama, which he created and managed with Charles Marie Bouton (1781–1853) and Hippolyte Sébron (1801–79).[12] The Diorama was a scenic entertainment focused on a large painting created using translucent and opaque pigments on the front and back of the canvas. Hung in a specially designed viewing space, a Diorama painting would present an initial effect, and then appear to change through the manipulation of light on the front and back.[13] The audience viewed the images from a darkened auditorium designed to heighten the illusion of the lighting effects. A show typically included two different paintings, an interior scene and an exterior, each hung in a separate room with its own lighting effects. Each scene created a theatrical *mise-en-scène* without actors, prompting the audience to participate imaginatively in the space.[14]

The Diorama drew from several precedents, including gardens such as the Jardin Monceau in Paris, where illusionary perspectives were juxtaposed with actual garden spaces.[15] Several anecdotes about the Diorama imply that viewers believed the illusion. Mentienne writes that during a show, Charles X (1757–1856) threw a two-franc piece at the image, expecting it to land in the depicted garden.[16] In light of the precedents set at the Jardin Monceau and similar spaces where illusion and real space are presented, the expectation that the Diorama image may, in fact, just be a window to an outside space becomes a real possibility. Visitors wanted to enter the space, not believing that such illusions could possibly be painted on a flat surface.

In Bry-sur-Marne, Daguerre continued to use the techniques he developed in the Diorama, most notably in creating an illusionary expansion of the local church, l'Église Saint-Gervais et Saint-Protais.[17] Although the credit for the idea differs depending on the source, there is evidence that Daguerre and Mlle de Rigny wanted to turn the church into a cathedral.[18] Given the size of the village and the space of the church, this was not feasible. Rather than an architectural expansion, Daguerre drew on his experiences with the Diorama to visually expand the space. To achieve this, the apse of the church was extended to accommodate a window and space behind the altar to allow natural light to enter behind a Diorama-style painting[19] (Plate 1).

Inaugurated on 19 June 1842, the feast day of the church's titular saints, the Diorama painting depicts the interior of a gothic-style cathedral, complete with groin vaults, second-storey gallery windows, evidence of a transept in the middle ground, and an apse in the background.[20] Windows appear to pierce the side aisles, illuminating the nave. In the foreground, the entrance to the painted cathedral is framed by compound pillars, hung with four frames, three of which contain paintings. The lintel connecting the pillars is decorated with gothic-style tracery, from which hangs a crucifix. Above the lintel is an empty cross signifying the resurrection.[21] There are painted curtains on the inside of the pillars, indicating that the scene has just been opened to the viewer. The painting continues beyond the pillars, heightening the illusion that one is standing inside a gothic cathedral. Due to the painting techniques and lighting effects, the light in the painting appears to move; shadows shift, candles appear to light and are snuffed out, the quality of the light changes, dimming or becoming more brilliant in concert with the natural shifts of sunlight outside. Since its restoration in 2012, these effects are no longer created by natural light in an effort to preserve the painting. Artificial lights were installed to recreate the effects in a shortened timespan for today's visitors.[22]

These effects are subtle, but clearly had a considerable impact on visitors throughout the nineteenth century. Although some disputed the suitability of such a spectacle in a church, it was mostly lauded as a success in its own time.[23] In the 1880s, M. Balagny presented his thoughts on visiting Bry-sur-Marne to the Société Française de Photographie, an account of which was published both in the Bulletin for the Society and by Gaston Tissadier in *La Nature*. Balagny begins his account with a detailed description of the church Diorama painting.[24] He recounts the placement of the Diorama painting behind the altar with the tabernacle, a specially designed container that holds the remains of communion after Mass, at the centre near the bottom of the painting.[25] The placement of the tabernacle, although typical for a Catholic church, adds layers of meaning when considered in relation to the painting. The tabernacle creates a visual and physical link between the actual church and the painted church, which reinforces the religious symbolism of communion as an act that joins the worldly, the body, with the divine represented by the imagined cathedral. In this light, the cathedral becomes more than a cathedral – it becomes a representation of the heavenly House of God. The painting thus participates in the Mass and the rituals central to the church itself, creating a liminal space that allows for a connection and crossing between physical and invisible realities.

Beyond being highly illusionistic, the painting and its placement creates a space that has a real effect on the reality and life of the town through the evocation of the cathedral.[26] Such distinctions were also made in the nineteenth century. Balagny's account commends the illusion, and notes that the painting 'makes the spectator believe that the altar is only in the middle of the church, and that behind it begins a new church, much more beautiful than the modest church of Bry'.[27] In describing the painting and its placement, he notes that it creates a 'necessary junction between the real church and the imaginary'.[28] The real and the imaginary meet in the space at the altar. Given the religious context, the 'imaginary' has metaphysical as well as aesthetic implications. The imagined cathedral extends the physical church into the cathedral while reinforcing a connection between the physical and the divine. As he was publicly affiliated with Roman Catholicism, and specifically the Catholic church at the heart of Bry-sur-Marne, Daguerre was likely sensitive to the liturgical potential of the imagery and its spectacular qualities.[29] Mlle de Rigny's involvement also highlights her support of and participation in the spiritual life of the village.[30] A few years after Daguerre's death, a speech given at his memorial service in Bry-sur-Marne highlighted this transformational quality, but in a more secular manner by calling

Daguerre the magician who had transformed the church into a cathedral.[31] This translates the religious experience into a magical one, where Daguerre becomes the magician through which the viewer is able to interact with the cathedral.

As it did in the case of the Bry-sur-Marne church, the Diorama framed much of Daguerre's work throughout his time in the town. Daguerre's Diorama experiments continued alongside his photographic work at his *maison de campagne*, which included a lab and a garden.[32] Having drawn, painted and photographed picturesque scenery throughout his career, Daguerre began implementing some of those characteristics in his own home. He created what Mentienne later describes as a Swiss-style landscape, a term often associated at that time with the elements of picturesque naturalism.[33] Although fairly large for the time, Daguerre's house garden did not have the space for such sprawling vistas. No matter, as with the church Daguerre did not need physical space. Instead, he 'elongated his park as he elongated the church'.[34] This implies that Daguerre included perspectival devices in his garden to complete the naturalistic illusion. Such illusions were well known and used in landscape gardens large and small for centuries before Daguerre.[35] After the success of the Diorama and other such entertainments, however, such devices were explicitly linked with these spectacles and their attendant connotations of illusion and theatricality.

Much of the scholarship on Daguerre's time in Bry-sur-Marne has neglected to differentiate this house garden from the landscape created at the local château. A similarity in aesthetic and relative lack of physical evidence, as well as the disappearance of both spaces, has added to the confusion. Yet Daguerre's involvement in the château landscape is significant, as its timing and the involvement of the citizens of the town reorient the implications of the landscape from the personal to the social and public. As the property of Mlle de Rigny, the château estate is a private space, and as the former *seigneurie* it is also part of the historical and symbolic fabric of the town. Unfortunately, due to the events of the Franco-Prussian war in 1870 which caused the burning and destruction of the grounds and château, along with most of its contents, there are few archival records to indicate specific access rights Mlle de Rigny may have granted to visit the landscape.[36] Ideas of public and private were very much in flux during the late eighteenth and early nineteenth centuries in France as the political climates and attitudes toward land rights and use changed. Generally speaking, however, access to such landscapes was determined by owners, making them private, yet their funds and lands made up the fabric of public and political life.[37] Daguerre's involvement in this specific landscape was, in fact, a direct result of this political and social unrest.

In 1848, an economic crisis in France resulted in a steep surge in unemployment and rising discontent. In February of that year, Louis-Philippe abdicated, ending the Orléanist Restoration government. The short-lived Second Republic that followed developed the *Ateliers Nationaux*, or national studios, as an emergency relief programme to employ citizens. The sudden removal of this programme in June of that same year led to a violent though brief civil war in Paris. This revolution ended with the election of Louis-Napoleon Bonaparte as President of the Second Republic, who would become Napoléon III under the Second Empire in 1851. According to Mentienne, it was in response to this socio-political climate that Mlle de Rigny commissioned Daguerre to make several changes to the château's landscape:

> In 1848, during the Revolution, this Lady, who had seen 1793, became frightened. Mr. Mentienne advised her to occupy as many workers as possible; following this advice she commissioned them to establish worksites on her property; so Daguerre organized national studios in the Bry park, and there made changes in the landscape.[38]

Although Mentienne gives his father, Armand-Louis Mentienne, credit for the idea of occupying the workers, it is clear from the framing of the event that Mlle de Rigny is the impetus behind the project, while Daguerre is credited as the artist who organizes both the design and the workers. The inclusion of the Mayor Mentienne and Daguerre, as *conseiller*, in the planning and improvement of the estate, as well as the employment of approximately twenty local workers to execute those designs, situates the château landscape somewhere between private estate and public project.[39] The landscape thus functioned as a retreat for de Rigny and the privileged bourgeoisie from the social and political disruptions of that year, as well as a reprieve for the workers who faced economic hardship up to that point.[40]

As it was known in Daguerre's time, the château at Bry-sur-Marne dates to 1760, when Étienne de Silhouette (1709–67) acquired the château and property.[41] It was originally thought to be the work of Jean-Baptiste Bullet, but has since been attributed to François Il Franque (1710–93) under Silhouette's patronage.[42] During that time, Silhouette also expanded the landscape, and left the entirety to his cousin and close friend, Clément de Laage, whose arms appear on the 1787 *Atlas de la Seigneurie de Bry* (Figure 2.1).[43] The *Atlas* shows the château and gardens implemented under Silhouette and de Laage in a classically symmetrical style, with geometric gardens framing an *allée* that extends from the town through the centre of the château and into the landscape beyond. The entirety exemplifies the so-called 'French' or formal style, of which Versailles is the largest and best-known example.

During the revolution of 1789 the grounds and castle were seized and eventually reacquired by Clément de Laage's son Philippe.[44] Due to financial constraints, however, Philippe de Laage rented the château to various tenants during the early 1800s, including Charles-Maurice de Talleyrand-Périgord (1754–1838). It is through Talleyrand that Baron Louis became familiar with the property. Baron (Abbé) Louis earned the title of Baron of the Empire in 1810, and subsequently supported the Restoration of the Bourbon

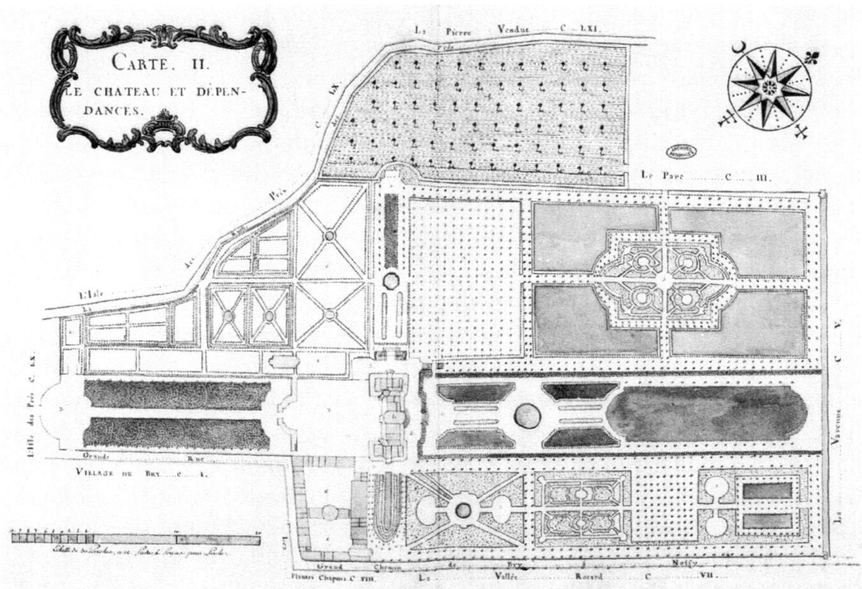

FIGURE 2.1 Communes du Département de la Seine, *Carte II Le Château et Dépendances,* from the *Atlas de la Terre et Seigneurie de Bry,* 1787. Archives Nationales de Paris, CP N IV Seine 1.

Monarchy. In 1814–15 he served as the minister of finance, and served again in the same position under Louis-Philippe (r. 1830–48).[45] He purchased the château and estate in 1816, which was thereafter inherited by Mlle de Rigny.[46] Unfortunately little is known about Geneviève de Rigny beyond what Mentienne includes in his accounts, where he notes her generosity and a keen interest in the arts and sciences.[47]

A map dated 1819 included in Vincent Roblin's 2012 history of Bry-sur-Marne indicates that the landscape was beginning to take on a less formal quality under the ownership of Baron Louis, though the geometric gardens nearest the château remained intact.[48] A survey of the town from 1840 shows the castle much as it appears in the 1787 map, and denotes a much larger area called 'Le Parc', noted in the accompanying ledger as belonging to Mlle de Rigny.[49] The events of 1848 appear in Mentienne's account as the precipitating factor in Mlle de Rigny's decision to alter the château landscape. As an *Ateliers Nationaux* project, the design and implementation of the landscape garden was a direct response to the political upheavals of that year. The choice to continue developing the informal aesthetic, as well as the socio-political moment, are instructive here.

The few records we have of the resulting landscape classified it as natural and 'picturesque'.[50] The picturesque has long been the focus of aesthetic debates. Its definitions and implications have been thoroughly discussed and debated since the eighteenth century in treatises on aesthetics and landscape gardens.[51] By the late 1700s the term was widespread to the point of being meaningless.[52] Generally meaning picture-like, the aesthetic emphasized design concepts found in landscape painting. Picturesque landscapes emphasize variety and naturalistic forms, such as winding paths and clumps of trees, in a distinct departure from the formal geometry of the so-called French style. Such naturalistic elements were often enhanced with architectural and sculptural works, usually ruins, rustic bridges and/or cottages, and sculptures or other material references drawn from classical antiquity.[53]

The difficulty in pinning down the particulars of the picturesque are clear from the vast number of treatises in the eighteenth and nineteenth centuries attempting to do just that.[54] In pre-revolutionary France, the term often conjured English or *jardin anglo-chinois* associations. Such associations may have appealed to Baron Louis given that he spent time in Britain during part of the French Revolution.[55] This style was translated, both in text and landscape, to France, adding further layers of complexity to its definition and associations.[56] It was championed by some as a uniquely French development, most notably Louis Carrogis Carmontelle (1717–1806) in his work on the Jardin Monceau.[57] Weltman-Aron argues that Jean-Marie Morel's (1728–1810) *Théorie des Jardins* of 1777 identifies the naturalistic style associated with the picturesque with universal associations of liberty, rather than specific political connotations.[58] Morel argues that the countryside, in particular, is the 'theatre of [Nature's] magnificence and liberty', which is admired 'in all countries and in all centuries'.[59] The development of Ermenonville by René de Girardin (1735–1808) and its explicit associations with Jean-Jacques Rousseau (1712–78) further charged the aesthetic with philosophical qualities.[60]

While several scholars since then have identified particular political implications in specific theoretical writings and designs, the picturesque as a whole tends to defy or deflect any neatly circumscribed association.[61] Susan Taylor-Leduc's recent work on the picturesque during the late eighteenth to mid-nineteenth centuries demonstrates its use by both royal and Napoleonic patrons.[62] Nicholas Green argues that between the 1789 revolution and the Second Empire, the picturesque does not have a codified political or even class association, but rather is indicative of *natura naturans*, nature as herself, as part of a new metropolitan ideology.[63]

Daguerre was well aware of the qualities of the picturesque and its popularity, as he was one of many contributors to Charles Nodier's *Voyages pittoresques et romantiques dans l'ancienne France* (1820–78). He contributed two depictions of religious architecture, one an exterior view and the other an interior. The exterior view he provided for Nodier's project was of an abbey in ruins with vegetation growing around the crumbling edges, art succumbing to the forces of time and nature.[64] Figures at the bottom offer scale as well as an imaginative invitation to the viewer. The other is an interior view of a church in a gothic style.[65] These types of views, along with the pairing of interior and exterior scenes, became expected at the Diorama and in many ways the aesthetics of the picturesque merged with the illusionism of the Diorama itself.

As Stephen Christopher Pinson notes, Daguerre's knowledge of the picturesque aesthetic comes full circle in his designs for the château landscape.[66] A map from 1859 (Figure 2.2), which shows the pieces of the estate that were for sale after Mlle de Rigny's death, gives us some indication of the landscape design after Daguerre's involvement. The map shows the sale of the main château and bulk of the landscape to a Parisian chocolatier, François Jules Devinck.[67] During Daguerre's time, the estate extended from this parcel of land noted as 'Lot du Château vendu à M. Devinck' west almost to the banks of the Marne river and north to the edge of the town at the top right of the image. Under the lot markers and blue highlights are faint lines indicating landscape features, including winding paths and groves of trees. The formal plantings of De Laage's time are barely detectable just to the south of the château. The rest has been thoroughly redesigned into a more naturalistic landscape. The sinuous paths and a small river wind through the landscape, coming together at the northern edge of the property in a large clump of trees and a small lake. This area, marked as lot 252 on the map, identifies this oasis as the *Grotte*, or grotto, a reference to one of the most defining features of Daguerre's work on the property.[68] Focusing this northern corner of the estate on a grotto and a water feature prefigures some of the notable qualities in the later landscapes of Buttes Chaumont and Bois de Vincennes.[69]

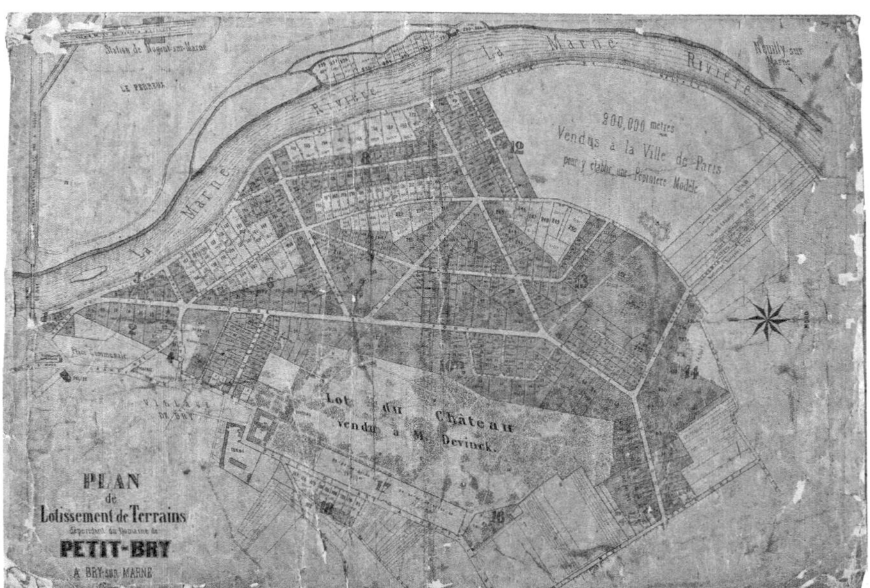

FIGURE 2.2 *Plan de lotissement du domaine de Bry*, 1859, Archives municipales de Bry-sur-Marne, 1Fi11.

Mentienne's account focuses on the grotto, with a brief description of some of the other features to develop an overall sense of the landscape. He mentions 'a perspective of a ruined castle in the distance that appears around the bend in the path', as well as the remains of a chapel, recalling Daguerre's other religious work in his church Diorama.[70] Whether this ruined castle is another perspectival device or an architectural *fabrique* is unclear. To these architectural features he adds that Daguerre's design included a miniature lake, boulders, rustic bridges and fir trees. There were plans to add a waterfall that would flow from the ruins of the chapel and into the lake, presumably the one in front of the grotto. However, both Mlle de Rigny and Daguerre died before the waterfall was completed.[71] We could identify the picturesque aesthetic from these qualities, but Mentienne also describes it as such: 'This landscape had a very picturesque quality, and the changes in the landscape were so well-executed that one might take them for natural.'[72] The turn of phrase at the end is telling: the landscape was so well designed that one might take, or accept, it for nature itself. The artifice here is emphasized as a key quality of the landscape; it is a translation through the work of Daguerre into a spectacular form of naturalism.

This description of the landscape very closely matches a photographic reproduction by Mentienne of a drawing attributed to Daguerre (Figure 2.3). In the foreground of the photograph is the shore of a lake or river with a path to the right that invites the viewer into the landscape. The middle ground is occupied by a grotto, formed by a rustic stone arcade, along with a rustic bridge and ruins on a hill. These are framed by trees to either side with a

FIGURE 2.3 Louis-Jacques Mandé Daguerre (drawing), Adrien Mentienne (photograph), *Grotte et ruines dans le Parc de Bry du temps de Mlle. De Rigny* (Grotto and ruins in the Bry Park during the time of Mlle de Rigny) c. 1848–50, printed later. Gelatin silver print, 16.2 × 33.5 cm. Gift of Eastman Kodak Company: ex-collection of Gabriel Cromer.

hazy, undefined background. Inside the grotto, under the middle arch, a barely discernible figure draws the viewer's eye into the shadowy depths overlooking the water. The arcade creates a permeability to the structure, inviting the viewer to explore its depths, further enforced by the presence of the figure.

The relationship of this image to the actual landscape is ambiguous. It is also a photographic reproduction of a drawing, thereby introducing a further layer of translation. The photograph includes an inscription on the back by Adrien Mentienne: 'Grotto and ruins in the Bry park in the time of Mlle de Rigny, made by Daguerre, circa 1850. Taken after a sketch by Daguerre inventor of photography and the diorama.'[73] If the attribution of the original drawing is correct, it could be a planning sketch, an image that served as source material, or an imaginative sketch of the elements as they existed in the landscape itself. The inclusion of the ruins, which may have been part of a perspectival device rather than an actual architectural feature, demonstrates a slippage in the image between the physical landscape and the imagined extension. The panoramic quality of the work, providing a view of all of the elements mentioned by Mentienne, as well as the collapsing of imagined and actual features indicates that this is most likely a planning or source sketch for the landscape designs.

Although the grotto and landscape no longer exist, there is evidence that it was maintained through the early twentieth century. Tissadier's article about Balagny's visit to Bry-sur-Marne in 1889 includes a print labelled 'Grotto built by Daguerre' that features the grotto as seen from across the artificial lake with the rustic bridge. Two of the three arcades described in Daguerre's sketch are shown with a human figure under one of the arches. The rusticated stone of the grotto is obscured by trees and vegetation. A 1912 photograph titled *En Famille, Près de la Grotte de Daguerre*, or 'The Family, near Daguerre's Grotto', shows a similar scene and provides further evidence that the grotto remained in a similar state through the early twentieth century (Figure 2.4). The photograph features the grotto with its three rustic arches, very much like the one in the drawing attributed to Daguerre. To the viewer's right a young child sits precariously on a bridge, while the rest of the family occupy positions around the arches, including one adventurous youngster on the top of the grotto. Shadows and foliage obscure the left side, while the bridge and path indicated on the right are separated from the viewer by the lake.

The figures in the photograph from 1912 provide a good idea of these features' proportions, while their positions throughout the structure indicate that they were interactive. An undated drawing from the same collection titled *Grotte de Daguerre* by Charles Lantz (Figure 2.5) further clarifies several features.[74] Lantz's drawing focuses on the lake and the nearby bridge and grotto with a perspective and light source very similar to that seen in the 1912 photograph. To the viewer's right a path leads from the grotto over a wooden bridge spanning a small river, then disappears behind a large tree,

FIGURE 2.4 *En Famille, près de la Grotte de Daguerre* (The family, near Daguerre's grotto), 1912, Raymond Lantz Collection, Ville de Bry-sur-Marne/Musée Adrien Mentienne.

FIGURE 2.5 Charles Lantz, *Grotte de Daguerre* (Daguerre's Grotto), n.d., Raymond Lantz Collection, Ville de Bry-sur-Marne/Musée Adrien Mentienne.

the contrast and bulk of which draws the eye into the foreground. Similar to Daguerre's drawing, the Lantz sketch provides some indication of a shore in the foreground, a place for the viewer to imagine standing. The drawing indicates characteristically picturesque moments, such as the partially dead tree and fallen log in the foreground, as well as the rustic wooden bridge and crumbling arches. Such features are common in picturesque landscape painting as they offer visual variety, but given the framing of the landscape in terms of Daguerre's work in the church could also imply a more religious reflection on death and resurrection. Whether religious or purely aesthetic, the artist has drawn the picturesque landscape according to established codes of picturesque looking and representation.

Behind these main features, the background is cut off from view by foliage. The dense wall of trees creates a backdrop that prevents us from seeing beyond the scene in both the Lantz drawing and the *En Famille* photograph. Unlike the Daguerre drawing's more open depiction, these works imply that there is no view out into the surrounding landscape. This little oasis, as Balagny described it, appears to exist in a hollow or bordered by thick forest or groves. As Mentienne laments, the trees had grown to the point of ruining the panoramic effect of the original design that visually connected the grotto with the rest of the former château landscape.[75]

The most enigmatic image of the grotto is an undated postcard, the origins of which are unclear (Figure 2.6). The focus is again on the grotto arches seen from across the lake, but from a different angle than that shown in the Lantz drawing and photograph. Shadows seem to form a face that peers out from the darkened archway. The features are partially obscured, but the proportions, given the size of the archway as indicated in the 1912 photograph, would make the face as large as a fully grown adult. It is unclear whether it is an apparition, a folly that actually existed, or the postcard artist's own whimsy, perhaps responding to Daguerre's reputation as a 'magician' of images.

The caption is equally ambiguous: '*Bry-sur-Marne* – La Grotte où Daguerre aimait à se reposer de ses travaux', 'the Grotto where Daguerre liked to rest from his work'. This focuses the space on Daguerre's presence without specifying his role in the grotto. The most explicit implication is that this is the garden where Daguerre would not work, but merely a place for him to visit and enjoy. The conception of nature and the countryside as a space where Parisians of means would rest, away from the city, has been well documented and analysed by Nicholas Green.[76] Mentienne notes that Bry-sur-Marne had become a retreat for Daguerre after the destruction of his first Diorama building in Paris, which burned down in 1839.[77] The postcard further focuses Daguerre's retreat to the specific site of the grotto.

Digging deeper into the rhetorical turn of phrase, 'to rest' also has biblical implications. Given the illusionism credited to Daguerre's Dioramas and photographs, this might insinuate his creation of fictive realities, as the magician of the cathedral, and this garden is where he took his rest, much like

FIGURE 2.6 *La grotte où Daguerre aimait à se reposer de ses travaux* (The grotto where Daguerre loved to rest from his work), postcard, Collection Musée Adrien Mentienne. Photo credit © Musée Adrien Mentienne, Ville de Bry-sur-Marne.

the Judeo-Christian God in the Garden of Eden. Given Daguerre and Mlle de Rigny's associations with Catholicism and the village church, the biblical tone here suggests a religious quality to the landscape they created. That association aligns the grotto and ruined castle with a sacred quality related to Catholic concepts of mortality and the passage of time, as well as a site of contemplation and prayer. Whether these beliefs are at the heart of the landscape's conception is unclear, but certainly the pairing of the landscape with the church Diorama would imply or strengthen such associations.

Whether specifically Catholic or not, Edenic analogies were common in garden spaces, and the inclusion of ruins and grottos suggests ruminations on the passage of time and the power of nature over the works of man and empires.[78] Grottoes in Europe by the eighteenth century often connoted hermitages, rustic retreats that provided space for contemplation.[79] Given the Catholic connection with the church Diorama, the grotto could be seen as a retreat from the material world, following the examples of notable Catholic saints.[80] Further, the phrase '*se reposer*', to rest, has melancholic as well as religious tones in its similarity to the phrase 'laid to rest', potentially signalling the artist's death. By pointing to Daguerre's presence in the garden, this '*se reposer*' also indicates his absence, his death, which effectively halted the completion of the garden space.

Combined with Mentienne's and Balagny's accounts and the 1859 map, these images create a fragmentary window into this landscape. The château is the focal point in the southern part of the property, nearest the town, while the grotto becomes its pendant anchoring the northern edges of the landscape. Along the path to the grotto there may have been a distant view, or a perspectival device, of a ruined castle. In Balagny's description there is an indication that the Marne river may have been visible from the grotto, an observation supported by the placement of the lake and foliage of the grotto on the 1859 map.[81] A viewer standing inside the grotto looking out may have had a picturesque view of the surrounding landscape itself. The combination of the darkened interior with the panoramic view of the Marne evokes the Diorama itself with its darkened audience room and picturesque views illuminated by natural and artificial light. The grotto becomes both a picturesque feature and focal point of the landscape, as well as a place that frames the surrounding landscape as another picturesque Diorama.

Mentienne uses the Diorama concept to frame his entire description of the landscape with the phrase 'natural diorama'.[82] On the surface, this may allude to the use of the perspectival devices in the landscape. But this turn of phrase, foregrounding the whole description, complicates the relationship between nature and its representation. The rhetorical slippage involving both the Diorama and this landscape garden revolves around a telling juxtaposition between nature and artifice. Describing the landscape as a natural Diorama implies that the spectacle was not natural, as opposed to the natural scenery of the landscape. In the most literal sense, this is true; the Diorama is an illusion, while the landscape is a physical space made of real, that is, physically present and interactive, materials. Yet the experience of the Diorama was also characterized as 'nature itself'.[83] To complicate this further, Mentienne's description of the landscape describes it as a natural, or physical, realization of one of Daguerre's Diorama scenes. In this, the landscape is the copy and the Diorama is the original. Pinson addresses this issue of the copy and original in an analysis of a review of the Diorama in which the reviewer objects to the Diorama's attempts to make the spectator 'accept a copy for the original'.[84] Pinson argues that this is indicative of a so-called counterproof phenomenon sensed in Daguerre's Diorama, distinguishing the spectacle from traditional painting.[85] In printmaking, a counterproof is a second copy made from a print that is often used to inspect the original plate as it is in the same state, making it closer to the original than the prints themselves. The counterproof, Pinson argues, rests on a paradoxical relationship between original and copy, between the authentic and the illusion. The reviewer's comments therefore reveal an anxiety about the spectacle's ability to confuse nature and artifice.[86]

In the château landscape at Bry-sur-Marne Daguerre creates a real space using natural or picturesque scenes from the Diorama as the source. Like a counterproof, the copy is used to shape and recreate the original scene in the landscape. Inverting illusion and nature, representation and original, the implication is that Mlle de Rigny's landscape becomes a type of Diorama through Daguerre's interventions. The aesthetic qualities reinforce this connection, as the type of picturesque nature that viewers had been conditioned to expect in artworks and shows like the Diorama take shape in the physical landscape. The natural aesthetic recalls and reinforces the slippage between the real and the illusion. While the Diorama was a theatrical set without actors, its effects and illusionism invited the audience to become the actors. This requires the viewer to recognize their participation in the scene, to internalize it and make it real in the imagination.[87] The desire to physically interact with the type of spaces typically viewed in the Diorama becomes a reality in the landscape. Daguerre's Diorama paintings become the original which the artist recreates or imitates in the château landscape.

The Diorama, then, explicitly frames all of Daguerre's projects in the town: the church, his home, and the château landscape. By virtue of its placement and use, the church Diorama provides a fictive space, an illusion that is true. The illusion becomes an object of admiration, engagement and participation only once it is acknowledged as such. In that engagement, the viewers enter into a dialogue with the work, allowing themselves to be fooled and at the same time seeking to establish the fiction as artificial. By entering into this dialogue, the illusion becomes part of the mental process in which the visitor engages, becoming real by its participation in the space in the mind of the viewer.

In the case of the château landscape, the Diorama in general, and the church Diorama specifically, frames the design and reception of the space. In the uncertainty of 1848, the landscape is transformed through the labour of the local workers who manifest the real from the imagined. The aesthetic reference to the Diorama highlights the involvement of the community in making the imagined into a reality. According to Green, picturesque parks and gardens connoted a moral and physical good for all classes and ages during the period, while the ruins and grotto allude to the passage of time.[88] The combination of these elements creates a potent statement that highlights the potential of the moment to create or destroy. Nature's abundance and liberty, according to Morel, is able to 'calm the violence of destructive and evil passions'.[89] The château landscape becomes a retreat from the revolution for Mlle de Rigny and those in her circle, while also occupying the local populace. The creation of the site extends an economic reprieve to the workers, while its aesthetic qualities leverage the connotations of abundance and liberty associated with nature. As a response to the political events of 1848, the château landscape is adamantly political, while simultaneously creating an escape for the people of Bry-sur-Marne from that same political reality. The pervasiveness and ambiguity of the picturesque aesthetic perfectly capture and capitalize on this paradoxical multiplicity of meaning.[90] As a citizen of Bry-sur-Marne, Mentienne's own account perfectly encapsulates the effectiveness of Daguerre's aesthetic interventions in the village.

As a natural Diorama, the landscape is both insistently nature and illusion. This recalls Baudelaire's comments in his review of the 1859 Salon, where he refers to the Diorama specifically after criticizing the landscape paintings that year for having neither natural beauties nor imagination.

> I wish to be brought back to the dioramas, where the brutal and enormous magic knew how to create a useful illusion. I prefer to contemplate some theatrical sets, where I find my dearest dreams artistically expressed and tragically concentrated. These things, because they are false, are infinitely closer to truth; whereas the majority of our landscape painters are liars, exactly because they have neglected to lie.[91]

Baudelaire recalls Daguerre's Diorama as capable of expressing his dearest dreams. Only a decade after Daguerre's work at the château, Baudelaire insists that those illusions of the Diorama are closer to the truth because they are false. The Salon in question predominantly featured works of the Barbizon and realist schools and focused too much, Baudelaire agues, on specific details and flora. Among these he notes Constant Troyon (1810–65) as the 'best example of skill without soul'.[92] In Baudelaire's estimation, the landscape painters at the Salon lacked theatrical illusion, and in their efforts to represent what is real have neglected what is true. The aesthetics of illusion, nature that is perfected through art, such as the inclusion of ruins and other such touches, and presented through an all-encompassing spectacle like the Diorama, is preferable in its approximation to the ideal than the mere representation of reality.

The illusion of the Diorama creates a fictional scene, but in its interaction with the viewer becomes a reality. The viewer's dreams manifest themselves in the spectacle, an illusory experience that is no less true for its falsity. Or, in Baudelaire's estimation, because it is false, and therefore relies on imagination, it is closer to the truth. The imagination, as in the church Diorama, becomes the site of the divine made manifest by the theatrical illusion. Such dreams can, in many ways, surpass reality, creating a preference for the spectacle over the real. The unfixed nature of the picturesque, as an illusion made manifest in the landscape, offers an aesthetic retreat, a refuge from the difficulties of the 1848 revolution as well as a way to involve the community through the employment of local workers. While the space remained privately held, the landscape nevertheless functions in the public sphere: as functional land, as employment, and as a symbol for the community. At a time when the political foundations of the country were unstable, resulting in the disengagement of the citizens of Bry-sur-Marne, the aesthetic of a more perfect nature made manifest offered a preferable illusion over the reality of the moment. Just as Baudelaire longs for an escape into the grand illusions of the past, Daguerre and Mlle de Rigny use the aesthetics of picturesque illusion to retreat from the political and social turmoil of 1848. To call the château landscape a natural Diorama reveals its ability to create an idealized realization of the true, a dream of nature made manifest by the citizens of Bry-sur-Marne.

Notes

Unless otherwise noted, all translations are mine.

1 Adrien Mentienne, *La Découverte de la Photographie en 1839* (Paris: Imprimerie Paul Dupont, 1892), 111. *'Il conçut l'exécution d'un diorama naturel . . . C'était un coin charmant, et qui montrait avec quelles savantes combinaisons Daguerre savait simuler la nature.'*

2 Mentienne, *La Découverte de la Photographie*, 110–11.

3 The term 'Diorama' is capitalized to distinguish it from the colloquial term used today to denote three-dimensional models.

4 One of the earliest applications of the term '*fabrique*' to garden design appears in Claude-Henri Watelet, *Essai sur les Jardins* (Paris: Imprimerie Prault, 1774); Monique Mosser, 'Rapporter le tableau sur le terrain: Fabrique et poétique du jardin pittoresque' in *Jardins Romantiques Françaises: Du Jardin des Lumières au Parc Romantique 1770–1840* (Musée de la Vie Romantique, accompanying an exhibition of the same name 8 March–17 July 2011), 33–68.

5 For more about Alphand, see A. (Adolphe) Alphand, 1817–91. *Les Promenades de Paris: Histoire, Description des Embellissements, Dépenses de Création et d'Entretien des Bois de Boulogne et de Vincennes, Champs-Élysées, Parcs, Squares, Boulevards, Places Plantées, Étude sur l'art des Jardins et Arboretum* (Paris: Connaissance et mémoires, 2002); Richard S. Hopkins, *Planning the Greenspaces of Nineteenth-Century Paris* (Place: Louisiana State University Press, 2015); Heath Massey Schenker, 'Parks and Politics During the Second Empire in Paris', in *Landscape Journal*, 14, no. 2 (Fall 1995): 201–19; Chiara Santini, *Adolphe Alphand et la Construction du Paysage de Paris* (Paris: Hermann, 2021).

6 Georges Potonniée, *Daguerre, Peintre et Décorateur* (Paris: Paul Montel, 1935); Helmut Gernsheim and Alison Gernsheim, *L. J. M. Daguerre: The History of the Diorama and the Daguerreotype* (New York: Dover Publications, Inc., 1968); Stephen Christopher Pinson, *Speculating Daguerre: Art and Enterprise in the Work of L. J. M. Daguerre* (Chicago and London: The University of Chicago Press, 2012).

7 R. Derek Wood, 'Daguerre and his Diorama in the 1830s: Some financial announcements', in *Photoresearcher (European Society for the History of Photography)*, 6 (1994/95/96): 35–40.

8 Vincent Roblin, *Bry-sur-Marne: Histoire et Patrimoine* (Maury Imprimeur SAS, Malesherbes, France, 2012), 92.

9 Mentienne, *La Découverte de la Photographie,* 108.

10 Vincent Roblin, *Bry-sur-Marne*, 85–6.

11 Mentienne, *La Découverte de la Photographie,* 111.

12 R. Derek Wood. 'The Diorama in Great Britain in the 1820s', in *The History of Photography,* 17, no. 3 (Autumn 1993): 284–95.

13 Louis-Jacques-Mandé Daguerre, *An Historical and Descriptive Account of the Daguerreotype and the Diorama* (London: Nutt, Bookseller, Fleet Street, 1839. Facsimile reprint. New York: Kraus, 1969), 81–3.

14 Guillaume le Gall, *La Peinture Mécanique: le Diorama de Daguerre,* 2013 (Mare & Martin, 2013), 6; Sophie Thomas, *Romanticism and Visuality: Fragments, History, Spectacle*, Routledge, 2008.

15 Pinson, *Speculating Daguerre*, 44.

16 Mentienne, *La Découverte de la Photographie,* 102.

17 Hereafter referred to as the church or l'Église.

18 Mentienne, *La Découverte de la Photographie,* 108.

19 See architectural diagram in Vincent Roblin, *Bry-sur-Marne,* 70.

20 Mentienne, *La Découverte de la Photographie,* 109.

21 Margaret Calvarin, Curator of the Musée Adrien Mentienne in Bry-sur-Marne, in discussion with the author, 28 May 2014.

22 Vincent Roblin, *Bry-sur-Marne,* 95–6.

23 J. B. Duchaine (Abbé), 'Diorama de M. Daguerre, appliqué à l'église de Bry-sur-Marne', in *L'Ami de la Religion*, 114, 3635 (28 July 1842): 191–2.

24 Gaston Tissadier, 'Souvenirs de Daguerre', in *La Nature* (28 September 1889): 285. M. Balagny, 'Souvenirs de Daguerre', *Bulletin de la Société Française de Photographie* (Paris, 1889): 212–15.

25 Tissadier, 'Souvenirs de Daguerre', 284.

26 Margaret Calvarin, in discussion with the author, 28 May 2014.

27 Balagny, 'Souvenirs de Daguerre', 213.

28 Ibid.

29 Jean Roblin, Philippe Drancy and Marieke Housseau, eds., *Hommage à Daguerre: Magicien de l'Image.* Introduction by Pierre Emmanuel, exhibition catalogue (Bry-sur-Marne, France: l'Hôtel de Malestroit, 1976), 15, 35. Daguerre was baptized in the Roman Catholic parish of Saint-Martin de Cormeilles-en-Parisis, and his funeral services were held at Saint-Gervais et Saint-Protais de Bry-sur-Marne.

30 Mentienne, *La Découverte de la Photographie*, 107–8.

31 Ibid., 118–19.

32 Pinson, *Speculating Daguerre,* 120. Daguerre's house and property was sold to the order of St. Clothild. The house was destroyed in 1870.

33 Mentienne, *La Découverte de la Photographie,* 105.

34 Balagny, 'Souvenirs de Daguerre', 215.

35 John Evelyn, *Elysium Britannica,* edited by John Ingram (University of Pennsylvania Press, 2000); John Dixon Hunt, *John Evelyn: A Life of Domesticity* (London: Reaktion Books, 2018), 52–3; Luke Morgan, 'The Early Modern "Trompe-l'Oeil" Garden', in *Garden History*, 33, no. 2 (Autumn 2005): 286–93.

36 Vincent Roblin, *Bry-sur-Marne,* 88.

37 Susan Taylor-Leduc, *Marie-Antoinette's Legacy: The Politics of French Garden Patronage and Picturesque Design, 1775–1867* (Amsterdam: Amsterdam University Press: 2022), 22.

38 Mentienne, *La Découverte de la Photographie*, 110. 'En 1848, au moment de la Révolution, cette dame, qui avait vu 1793, avait paru s'effrayer; M. Mentienne lui conseilla d'occuper le plus d'ouvriers possible; suivant ce conseil, elle les chargea tous deux d'établir des chantiers dans sa propriété; alors Daguerre organisa des ateliers nationaux dans le parc de Bry, en y faisant faire des mouvements de terrains . . .'

39 Vincent Roblin, *Bry-sur-Marne*, 92.

40 Mentienne, *La Découverte de la Photographie*, 123. One of the few sources that discusses access to the space during Mlle de Rigny's time. During the dedication of the Bry-sur-Marne memorial to Daguerre in 1851, Mlle de Rigny offered the hospitality of her château to those who had come from the Société libre des Beaux-Arts, as well as Daguerre's friends and colleagues.

41 Vincent Roblin, *Bry-sur-Marne,* 51.

42 Marie-Agnès Férault, 'Une œuvre inconnue de François Il Franque: le château de Bry-sur-Marne', in *Bulletin Monumental*, 149, no. 3 (1991): 299–315.

43 Vincent Roblin, *Bry-sur-Marne,* 55.

44 Ibid., 81.

45 Ibid., 83–4.

46 Ibid., 84–5.

47 Mentienne, *La Découverte de la Photographie,* 107–8.

48 Vincent Roblin, *Bry-sur-Marne,* 111.

49 Commune de Bry-sur-Marne, part of the Etats des Communes survey, Section A, Site du Château, 1840, Ville de Bry-sur-Marne/Musée Adrien Mentienne.

50 Mentienne, *La Découverte de la Photographie,* 111.

51 See Stephen Copley and Peter Garside, eds., *The Politics of the Picturesque: Literature, Landscape and Aesthetics since 1770* (Cambridge, MA: Cambridge University Press, 1994). See also Pinson, 60–1.

52 Copley and Garside, *The Politics of the Picturesque,* 1.

53 Ibid., 3. See also John Dixon Hunt, *The Picturesque Garden in Europe* (New York: Thames & Hudson, 2002), 10–11.

54 Some notable examples include Claude-Henri Watelet, Edmund Burke, Humphry Repton, Uvedale Price, Richard Payne Knight, Thomas Whately, Louis Carrogis Carmontelle, Jean-Marie Morel, René de Girardin, Pierre-Henri de Valenciennes, Alexandre de Laborde, etc. Hunt, *The Picturesque Garden in Europe*, 94–8.

55 Vincent Roblin, *Bry-sur-Marne,* 83.

56 Hunt, *The Picturesque Garden in Europe.*

57 Carmontelle, *Le Jardin Monceau, près de Paris, Appartenant à son Altesse Sérénissime Monseigneur le Duc de Chartres* (Paris: Chez M. Delafosse . . . Chez MM. Née & Masquelier . . . et chez tous les marchands d'estampes, 1779), 3.

58 Brigitte Weltman-Aron, *On Other Grounds: Landscape Gardening and Nationalism in Eighteenth-Century England and France* (New York: State University of New York Press, 2001), 65.

59 J. M. (Jean-Marie) Morel, *Théorie des Jardins* (Paris: Chez Pissot, 1776), 25–6.

60 Michael Charlesworth, *Landscape and Vision in Nineteenth-Century Britain and France* (Aldershot, England, and Burlington, VT: Ashgate, 2008), 82–4. See also Pinson, 60–1.

61 Copley and Garside, *The Politics of the Picturesque,* 6.

62 Taylor-Leduc, 17–18.

63 Nicholas Green, *The Spectacle of Nature: Landscape and Bourgeois Culture in Nineteenth-Century France* (Manchester, New York: Manchester University Press, 1990), 11.

64 Charles Nodier, Justin Taylor and Alphonse de Cailleux, *Voyages Pittoresques et Romantiques dans l'ancienne France* (Gide fils: Paris, 1820–1878), Plate 12.

65 Nodier, *Voyages Pittoresques,* plate 45.

66 Pinson, *Speculating Daguerre,* 120.

67 Vincent Roblin, *Bry-sur-Marne,* 61, 97–105.

68 *Plan des Lotissements* de 1859, lot 252, corresponding to the *Liste de acquirers du domaine de Bry de 1860* (Archives Municipals 1 O 39).

69 H. Parizot and A.V. Boileau, *Guide-Album Historique et Descriptif du Bois de Vincennes et du Chemin de Fer de Paris à Vincennes et à la Varenne-Saint-Maur* (Paris: M. H Parizot, 1860).

70 Mentienne, *La Découverte de la Photographie,* 111. 'Il conçut l'exécution d'un diorama naturel: c'était la perspective d'un vieux château fort en ruines dans l'éloignement.'

71 Ibid.

72 Ibid. 'Ce paysage avait un aspect très pittoresque, et les mouvements de terrains y étaient exécutés d'une façon si savant qu'on les prenait comme étant naturels.'

73 Louis-Jacques-Mandé Daguerre (attr. drawing), Adrien Mentienne (photograph), *Grotte et ruines dans le Parc de Bry du temps de Mlle. De Rigny* (Grotto and ruins in the Bry Park during the time of Mlle de Rigny) *c.* 1848–50, printed later. Gelatin silver print, 16.2 × 33.5 cm. Gift of Eastman Kodak Company: ex-collection of Gabriel Cromer.

74 Jean Roblin, Drancy and Housseau (eds), *Hommage à Daguerre,* 34–5.

75 Mentienne, *La Découverte de la Photographie,* 111.

76 Green, *The Spectacle of Nature,* 75.

77 Mentienne, *La Découverte de la Photographie,* 105.

78 Michel Baridon, 'Ruins as a Mental Construct', in *Journal of Garden History,* 5, no. 1 (1985): 94; Anne Janowitz, *England's Ruins: Poetic Purpose and the National Landscape* (Cambridge, MA: Basil Blackwell, Inc., 1990): 82.

79 Hervé Brunon and Monique Mosser, *L'Imaginaire des Grottes dans les Jardins Européens* (Paris: Éditions Hazan, 2014), 346.

80 Brunon and Mosser, 342–5.

81 Balagny, 'Souvenirs de Daguerre', 214.

82 Mentienne, *La Découverte de la Photographie,* 111.

83 *The Times* (London), 'Diorama', 4 October 1823, 3.

84 Pinson, *Speculating Daguerre,* 125.

85 Ibid.

86 Ibid.

87 Thomas, *Romanticism and Visuality,* chs 6 and 7.

88 Green, *The Spectacle of Nature,* 74.

89 Morel 24–5. 'La campagne, théâtre de sa [Nature's] magnificence & de sa liberté, offre à ses heureux habitans des ressources sans nombres: la vie y coule sans inquiétude & sans remords dans des occupations agréables & fructueuses; l'âme y est saine & le coeur en paix; son séjour calme la violence des passions destructives & malfaisants.'

90 Copley and Garside, *The Politics of the Picturesque,* 6.

91 Charles Baudelaire, *Curiosités Esthétiques: Salon 1845–1859* (Paris: Michel Lévy Frères, 1868), 338.

92 Ibid., 328.

PART II

Dwelling, garden and the poetic image

3

Three Settings from *Wen Hengshan's Album of Poems and Paintings of the Garden of Artless Administration*, 1533

Hongde Wu and Stanislaus Fung

Contemporary scholars of Chinese gardens have given detailed attention to Suzhou as a prominent centre of garden culture during the Ming dynasty (1368–1644). Many gardens were constructed during two major periods of activity in the Jiangnan area: 1465–1521 and 1522–1620.[1] The Zhuo Zheng Yuan or Garden of Artless Administration, popularly known as the Garden of the Unsuccessful Politician, was established around 1510–13, as part of what has come to be regarded as the first of two waves of garden-making activities.[2]

An album entitled *Wen Hengshan's Album of Poems and Paintings of the Garden of Artless Administration* was produced by Wen Zhengming (1470–1559) for the garden's owner, Wang Xianchen (who received the *jinshi* degree in 1493), in 1533.

The location of this album is unknown, and modern scholars generally rely on a photolithographic reproduction.[3] In the format that it has come down to us, the main body of the album comprises thirty-one double-page spreads, each portraying a scene from the Garden of Artless Administration with a poem in Wen's calligraphy and a corresponding painting by him. The introduction to each poem, written in prose, provides concise remarks on the location and characteristics of a setting. This is followed by a poem in regulated verse with four or eight lines. Thirty poems predating the paintings and calligraphy by two years likely had a significant influence on the conception of the paintings.[4]

Fifty years ago, the general context of the Garden of Artless Administration would have been considered in the following way. Private gardens served not only as sanctuaries but also as places for literary gatherings. Suzhou, home to over 200 gardens, attracted retired officials, literati and artists, fostering an elegant social environment.[5] Wen Zhengming, a prominent figure in the Wu School of Painting, played a central role in this social circle. His association with Wang Xianchen, a retired bureaucrat and owner of the Garden of Artless Administration, resulted in a variety of works inspired by the garden, including garden records, original poems and paintings.[6]

Has recent scholarship proposed new understandings of the cultural context of this garden? In a major contribution that aligned the history of gardens with a broader discussion of material culture and consumption in the Ming dynasty, Craig Clunas points to a major shift of status in which gardens as sites of agricultural production became sites of aesthetic enjoyment. In Clunas' estimation, 'half the land area of the garden' was taken up by 'a combination of flowering plums and apples in 1533'.[7] In this regard, the many plants of economic value mentioned by Wen Zhengming would be set aside in the subsequent aestheticization of the garden in historical accounts. The role of literary gatherings would overshadow the role of fruit trees.

It is not uncommon that contemporary Chinese scholars of gardens approach images with a naïve attitude towards 'representation', seeing in the paintings in Wen Zhengming's album the 'historical reality' of the Garden of Artless Administration. Recent scholarship, however, challenges this attitude. Liu Run'en argues that the textual and visual elements in the album do not always agree and should not be considered objective representations.[8] Wei Xiuyu suggests that many painting themes were inherited from other works of painting.[9] Lu Andong uses diagrams to illustrate the relative positions of various places in the album, implying a hidden garden route expressing a sense of a 'reclusive landscape'.[10] Through a detailed analysis of the text–image relationships in the album, Wen Zishen and Stanislaus Fung argue that the album as iconotext points to multisensory experiences, rather than objective reconstruction.[11]

This chapter will examine three settings from Wen Zhengming's album, explaining the development of garden experience. Building upon Emmanuel Alloa's critique of the 'transparency' presumption between language and image,[12] this study considers the differences in media between prose, poetry and images. It explores how they interact to form an aperspectival, embodied experience. In terms of understanding Wen Zhengming's poems in the album, we shall draw on François Cheng's diagrammatic analysis of the spatial-temporal structure of Chinese poems.[13]

Our basic premises can be summed up as follows. In contrast to the usual assumption of image and text as objective representation, we shall not be concerned with the 'reality' of the site and give priority to the ways in which a sense of 'latency' haunts the site.[14] Through a concrescence of different types of imaginings and experiences, we shall focus on how the album evokes a sense of material imagination and goes beyond a simple visual emphasis.[15] We shall also assume that poems and paintings go beyond one-to-one correspondence.

1. Storied Pavilion for Dreaming of Reclusion

The Chinese name of the pavilion Meng Yin (dreaming of reclusion) retrospectively figures the owner's dream as a foretelling, well before the owner bought the former home of famous recluses and retired there (Table 3.1).

Couplet 1, prefigured in a dream:

> Forests and springs enter a dream, with a vague sense.
> A storied pavilion rises up, as if in retreat and reclusive.

The first couplet reiterates the reason for the pavilion. It is a specific situation: the owner Wang Xianchen had dreamt of 'forests and springs' (*lin quan*, where recluses linger). The painting shows a person (probably the owner) in a storied pavilion. The curtains imply a setting for a bed (and also sleeping and dreaming). Outside is a dream-like landscape, and what intrigues us is the painted image's ability to indicate two readings. It might be taken to show the setting of the owner's dream at Jiu Li Lake, or, alternatively, the setting of the Storied Pavilion for Dreaming of Reclusion in the Garden of Artless Administration.

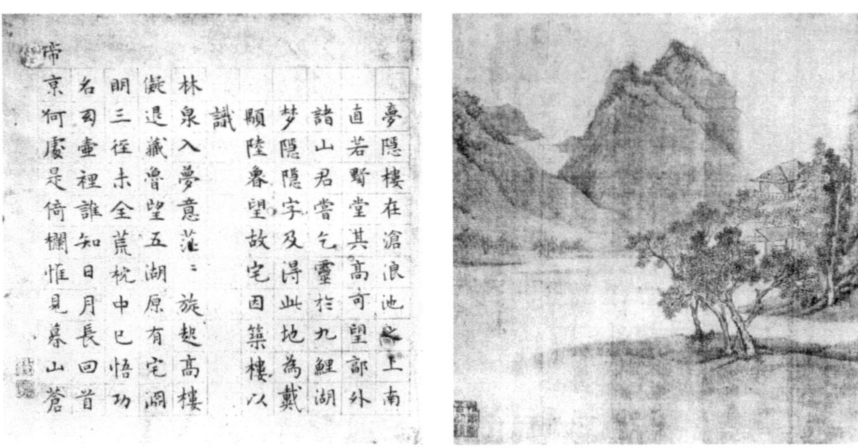

FIGURE 3.1 Wen Zhengming, *Meng Yin Lou*, or *Storied Pavilion for Dreaming of Reclusion*, 1533. Reproduced in Kate Kerby, *An Old Chinese Garden: A Three-fold Masterpiece of Poetry, Calligraphy and Painting by Wen Chen Ming, Famous Landscape Artist of the Ming Dynasty*. Studies written by Kate Kerby. Translations by Mo Zung Chung. Shanghai: Chung hwa Book Company, 1922.

Table 3.1 *Translation and transcription of Chinese text on 'Meng Yin Lou'.*

The Storied Pavilion for Dreaming of Reclusion looks over the Pond of Surging Waves, with the As-if-a-Rural-Retreat Hall immediately to its south. Its height allows a view of hills beyond the city wall. The Owner once consulted the omens at Jiu Li Lake, and, in a dream he had there, received the word '*yin*' (reclusion). Later, he bought this property, which was the former residence of Dai Yong and Lu Luwang, and he erected this storied pavilion to commemorate the dream.

Forests and springs enter a dream, with a vague sense.
A storied pavilion rises up, as if in retreat and reclusive.
Luwang originally had a residence here among the Five Lakes;
The Three Paths of Yuanming are not yet altogether decrepit.
In your dream, you had realized that merit and fame are illusory.
In a gourd, how can one count the days and months?
Looking back, where is the imperial capital?
Leaning against the balustrade, one sees only the sun setting on lush green hills.

夢隱樓在滄浪池之上，南直若墅堂，其高可望郭外諸山。君嘗乞靈於九鯉湖，夢隱'隱'字，及得此地，為戴顒 、陸魯望故宅，因築樓以識。

林泉入夢意茫茫，旋起高樓擬退藏。
魯望五湖原有宅，淵明三徑未全荒。
枕中已悟功名幻，壺裡誰知日月長。
回首帝京何處是，倚欄惟見暮山蒼。

Couplet 2, remembering and imagining:

The words 'A storied pavilion rises up' in the first couplet leads to the opening up of much more expansive, imagined scenes.

> Luwang originally had a residence here among the Five Lakes;
> The Three Paths of Yuanming are not yet altogether decrepit.

The second couplet refers to an imagined scene, the home of the recluse Lu Luwang (d. *c.* 881). The Garden of Artless Administration evokes memories of recluses' retreats: the old home of Lu Luwang's seems still standing here, and hints of Tao Yuanming's (*c.* 365–427) 'Three Paths' still seem to be 'around'. The new does not replace the old, it summons the old to present itself. Seeing and remembering are concurrent and correlated.

The storied pavilion in the painting, where we see a *pingzuo* balcony[16] on the upper level, is consistent with a public building rather than a building in a private garden. This reminds us that the painting is not a simple depiction of an actual scene.

In terms of basic format, the painting is consistent with a typical 'Painting of Dwelling in a Storied Pavilion' (*lou ju tu*).[17] The painter is quite proper in employing this format to evoke the environs of reclusion that might be associated with the hermits Luwang and Yuanming.

We can note here that the dynamic of then and now engaged by this second couplet is soon overtaken by a double sense of dreaming.

Couplet 3, self-reflection:

> In your dream, you had realized that merit and fame are illusory.
> In a gourd, how can one count the days and months?

'In your dream' is our translation for *zhen zhong*, incidentally also the title of a story from the Tang dynasty (*Zhen zhong ji,* Record within a Pillow)[18] in which a candidate for the imperial examinations fell asleep waiting for a meal to be brought to his table. The opening of the second line of this couplet refers to Hugong, an immortal known for miniaturizing himself and being able to dwell inside a gourd or calabash. These allusions to the stories of a mortal and an immortal analogize the owner of the garden. The moral of these analogic efforts – that merit and fame are illusory – is not presented in the mode of constative declaration but suspended, bracketed in the mode of a dream, as if one recognizes without being utterly clear, or as if one could only recognize it if it were not utterly clear, or as if recognition of the illusoriness is emergent and not simply available to a grasping mind. The painted scene, if it involves the dilation of temporal horizons that might make it tricky to keep track of the days and months, involves a shift of horizons on the part of the viewer of the painting.

Couplet 4, looking back and looking ahead:

> Looking back, where is the imperial capital?
> Leaning against the balustrade, one sees only the sun setting on lush green hills.

The first line of the fourth couplet about the capital that is 'not around' or beyond the horizon, echoes and contrasts with the scene of traces of previous recluses in couplet 2. The second line of couplet 4 brings to a close the sense of historic imagination and dreams in couplets 2 and 3, and returns us to the scene with the dying light of a landscape at sunset. The line does not distinguish the sense of landscape in the dilated temporal sense of being inside an immortal landscape from a scene in the Garden of Artless Administration.

The four couplets we have examined engage the reader in terms of imagined settings, retrospection and self-reflection by analogy. There is a peripatetic sense of moving between here and now and there and then,[19] in which perception and memory are woven together as dreamscape.

The painted scene also maintains the sense of an open work. The painting might be called 'Storied Pavilion for Dreaming of Reclusion in the Garden of Artless Administration' and equally 'Dream at Jiu Li Lake', or 'Former Residence of Luwang'. This openness of referentiality, by which we are engaged in the dreamscape, is released by text–image relationship and not simply a matter of stipulation. We would emphasize here the emergent sense of multiple meanings as we follow the indications and implications of the text–image relationship.

While the facing pages of poem and painting may not be a literal representation of a tangible landscape, they evoke the sensory richness of material imagination. They encompass not only the expansive and otherworldly qualities evoked by 'forests and springs', the five lakes, and a journey to the imperial capital, but also the immediate physical presence of a towering structure, the subtlety of meandering paths, and the sunset hues of a mountain vista. Thus, the allusion to the world within one's pillow and gourd serves as a hint to the tension between the cosmos and personal life.

2. Pavilion of the Pagoda Tree in Rain

Wen Zhengming's brief introduction to this poem ostensibly gives an account of the setting of this pavilion in the Garden of Artless Administration, but also indicates directly that the pavilion bears the sobriquet of the garden owner.

FIGURE 3.2 Wen Zhengming, *Huai Yu Ting*, or *Pavilion of the Pagoda Tree in Rain*, 1533. Reproduced in Kate Kerby (1922).

Table 3.2 *Translation and transcription of Chinese text on 'Huai Yu Ting'.*

The Pavilion of the Pagoda Tree in Rain is south of the Embankment of Peach Blossoms. To the west of the Pavilion is the Brook along Bamboos. Elms, pagoda tress, bamboos and cypress are all planted there. Huaiyu (Pagoda Tree in Rain) is the sobriquet that the Owner has adopted for himself.

The Pagoda Tree next to the Pavilion almost covers a wall;
Cool jade in the humid summer air moistens one's garments.
The sparse and gentle flowers of the Tree spread their fragrance afar;
Cool shade extending downwards benefits numerous generations.
Examinations in the Eight Month bring back memories;
The great careers of the Three Officials one leaves to one's descendents.
In old age, one would forego the dream of Nanke;
Moving the bed, one still reclines in the cool of the night.

槐雨亭在桃花浒之南，西臨竹澗，榆槐竹柏，所植非一。
云槐雨者，著君所自號也。

亭下高槐欲覆牆，氣蒸寒翠濕衣裳。
疏花靡靡流芳遠，清蔭垂垂世澤長。
八月文場懷往事，三公勳業付諸郎。
老來不作南柯夢，獨自移床臥晚涼。

Couplet 1, setting:

> The Pagoda Tree next to the Pavilion almost covers a wall;
> Cool jade in the humid summer air moistens one's garments.

The Pagoda Tree is experienced in terms of cooling shade. The shade of the tree reaches down while the humid summer air rises up, subtending a field.[20]

Couplet 2, atmosphere:

> The sparse and gentle flowers of the Tree spread their fragrance afar;
> Cool shade extending downwards benefits numerous generations.

In the fourth and fifth months of the lunar year, the fragrance of flowers spread far around the pavilion. In the summer months, and over many generations, the tree offers shade. This is likened to the ongoing benefits that descendants receive from their forebears. These benefits are a function of the good character and achievements of previous generations. The tree subtends a metaphoric, familial field.

Couplet 3, remembering and letting go:

> Examinations in the Eight Month bring back memories;
> The great careers of the Three Officials one leaves to one's descendants.

The first line of this couplet points to the eighth month of the lunar year, a time of the pursuit of official careers by sitting for examinations. The following line evokes a letting-go of a career, rhetorically, not as a matter of principle but as a matter of deferring to later generations. A traditional image would have the Three Officials seated in front of pagoda trees. Now in this pavilion, the conventional sense of this setting is displaced by a sense of letting go. Text and image, working now with the idea that the owner's sobriquet is at play, bring out the pivotal role of the solitary pagoda tree from the elms and cypresses, etc. Instead of showing a person in front of a pagoda tree, the tree is identified with the Owner.

Couplet 4, coming to equipoise:

> In old age, one would forego the dream of Nanke;
> Moving the bed, one still reclines in the cool of the night.

The opening line sets aside the dream of Nanke. In 'An Account of the Governor of the Southern Branch' (*Nan ke tai shou zhuan*),[21] Chunyu Fen dreamed that he married a princess of Huai'an, only to wake up and realize that in his dream he had entered an ant kingdom at the foot of a pagoda tree. When the rain came, he woke up next to the tree. The pagoda tree in the rain is an allusion to this story about a grand but illusory career. The final line refers to a wakeful owner moving his bed in search of a cool spot. We can imagine that he has now a cooler head to match a cooler setting. The poem ends with a quiet sense of open-endedness.

Much like the Pavilion for Dreaming of Reclusion, the Pavilion of the Pagoda Tree in Rain figures several places together: a dreamscape kingdom in miniature, a cool shaded setting with the smell of flowers in humid summer, the posture of a gentleman thinking of the contributions of an official career. It suggests putting oneself in a setting with multiple narrative dimensions, letting go of a previous path.

3. The Plantain Balustrade

In contrast to the previous two places of the garden we have examined, the Plantain Balustrade does not evoke memories or dreams but simply presents a place in the reader's imagination.

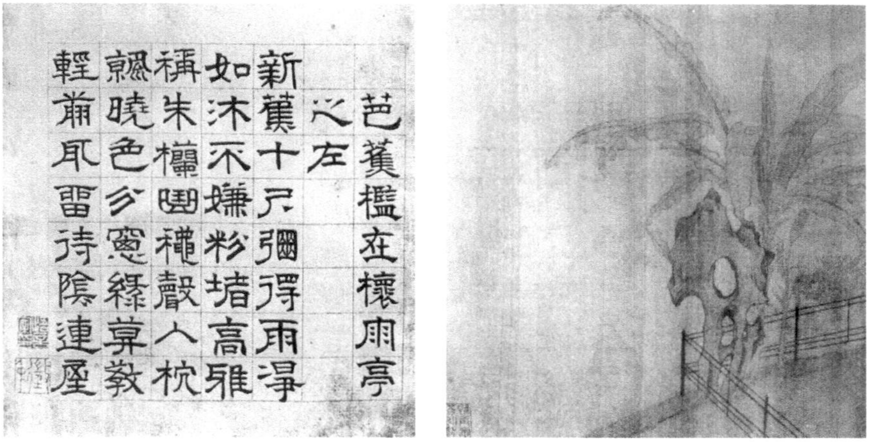

FIGURE 3.3 Wen Zhengming, *Ba Jiao Jian*, or *The Plantain Balustrade*, 1533. Reproduced in Kate Kerby (1922).

Table 3.3 *Translation and transcription of Chinese text on 'Ba Jiao Jian'.*

This tall plantain is to the left of the Pavilion of the Pagoda Tree in Rain.

The fresh plantain is over ten feet tall,
After a shower it is as clean as one after a bath.
It does not mind the high white wall,
And is well suited to the red balustrade that makes a zig-zag.
The sound of winds from the north reaches one's pillows in autumn.
Green shafts of light break up the window lattices at dawn.
Trim not the plant recklessly;
In time shade pervades the building.

芭蕉檻在槐雨亭之左。

新蕉十尺強，得雨淨如沐。
不嫌粉堵高，雅稱朱欄曲。
秋聲入枕飆，曉色分窗綠。
莫教輕剪取，留待陰連屋。

Couplet 1, the plantain in time:

> The fresh plantain is over ten feet tall.
> After the rain it is as clean as one after a bath.

This opening couplet delves into the tangible presence of the plantain upon initial observation, highlighting its impressive height and refreshing cleanliness following a shower. However, beneath the surface, three distinct temporal scales are subtly interwoven in these lines. The term 'fresh' subtly alludes to the ongoing year, considering that plantains are herbaceous and undergo a regrowth cycle in the subsequent spring. The height specification of 'ten feet' signifies the peak of summer, with the plantain on its way to overtake the eaves in height, entering maturity but not yet at its zenith. The phrase 'after the rain' pinpoints a specific moment, invoking sensuality by emphasizing the exceptional cleanliness of the plantain. These temporal nuances converge, creating an immersive 'now', or in Cheng's words, a 'closed time', as the beginning of the narrative.[22]

Couplet 2, the place where it stands:

> It does not mind the high white wall,
> And is well suited to the red balustrade that makes a zig-zag.

The second couplet meticulously fleshes out the backdrop against which the plantain exists, shedding light on its surroundings. Behind the tree, a tall white wall rises, while the red balustrade zig-zags along. These elements (white wall, the plantain and the red balustrade) come together to sustain a sense of place, employing intentional parallelism (*dui'ou,* a familiar Chinese rhetorical device in literary writings). They give the sense of a setting or a stage for a 'performance'.

A few further observations: firstly, unlike the symmetry found in poetic form, the elements depicted in the poem and those depicted in the painting do not form a strict one-to-one correspondence. For instance, the rocks featured prominently in the painting are omitted in the poem, hinting at a disparity between the two mediums. Secondly, unlike the static portrayal in the painting, the 'height' of the wall and the 'zig-zagging' of the rails in the poem operate dynamically, evoking a sense of 'becoming' rather than 'being'. This dynamic presentation guides the viewer's eyes beyond the painting, prompting the viewer to imagine encounters between the wall and the sky or more intricate variations in the railings, evoking a richer imagination than limited visual expressions can achieve.

Couplet 3, the atmosphere, expansion of the place:

> The sound of winds from the north reaches one's pillows in autumn.
> Green shafts of light break up the window lattices at dawn.

The third couplet offers a glimpse into an imaginary dwelling and its occupant. An expansion of space, extending beyond the picture, is achieved not by introducing new physical entities but by illustrating its non-physical effects. As the autumn wind sweeps through, the rustling sound of plantain leaves (making up the essential part of the 'autumnal sounds' in the preceding lines) is carried indoors, reaching a person lying down with a pillow. In the mornings, the leaves paint the windows with dancing shadows, imbued with a luminous hue, casting a serene and enchanting atmosphere. The poem adeptly evokes the atmosphere of the place, introducing dynamics such as autumn and morning. This dynamic portrayal aligns with Cheng's concept of dynamic space.

Couplet 4, advice on future actions:

> Trim not the plant recklessly;
> In time shade pervades the building.

In the final couplet, the poet wisely advises against reckless leaf trimming, suggesting a natural growth process until the shade generously covers the dwelling. The plantain's future is left in an unsettled state, an open-ended situation symbolizing open time. The dwelling implied in the previous two couplets becomes unequivocally evident. An expanding place is unveiled before the reader, merging with the shade of the plantain.

Intriguingly, the plantain, destined for pruning in winter, leaves behind a short chunk barely above the ground. The anticipated return of shade can only be fulfilled in the seasonal cycle of the next year.

The painting aligns visually with other works commonly referred to as 'The Early Spring' (*sui zhao tu*) and with paintings depicting specific seasons. By employing a material imagination of ephemeral elements such as flowers with fresh blossoms or plants with new shoots and rendering them in meticulous detail against a plain background, the painting effectively emphasizes sensual quality and latency within a particular moment.

The four couplets progress in layers, evoking a material imagination of a place dubbed 'the Plantain Balustrade' by introducing multidimensional sensations. In the opening couplet, one can sense the clean, moist quality of plantain leaves after rainfall. The following couplet juxtaposes the vibrant green of vegetation against the backdrop of white wall and red balustrade, creating a scene of contrasting hues. The third couplet introduces dynamic elements: the rustling sound of leaves stirred by autumnal winds, morning sunlight and green shadows playing next to the window, stirring imaginings beyond the painting. The closing couplet echoes the sentiments of the first couplet, emphasizing the yearly cycle of growth and advocating for an appreciation of a serendipitous coexistence shared.

Discussion

In the last 100 years, Chinese scholarship on the history of Chinese gardens has been often under the influence of naïve realism, the belief that one only has to open one's eyes to see the world as it really is, and that images are transparent to the world that they depict. In numerous publications on Chinese gardens from contemporary China, images and texts of diverse kinds and from all historical periods are mobilized for reconstruction of gardens.[23] The foregoing analysis of three settings from Wen Zhengming's album began with two uncommon leads. First, we drew on François Cheng's detailed account of Chinese poetry to tune into the nuances of Wen Zhengming's poems. This helped us understand how the modality of poetic expression might be related to understandings of the temporality of landscape experience. Second, we explored the implications of a previous study of multisensory perception in Wen's album by Wen Zishen and Stanislaus Fung.[24] In the course of developing our analysis, we encountered another dimension of the album in addition to the multisensory perception that displaced the privilege commonly given to the visual and the spatial in the reconstruction of Chinese gardens: memory, imagination and double readings are clearly involved in Wen's album.

Since our analysis appears at odds with the assumptions behind the agenda for reconstruction of gardens, we began to explore sources in image studies in the hope that our direction of thinking and the stakes involved in our work might be clarified. We were not hoping that specific publications in image studies might be entirely applicable to our study of Wen's album, and we should be wary of misleading correlations across cultural contexts and academic traditions. Our limited reading of publications in image studies also increases the risk of misunderstanding. We hope, however, that readers would take these notes as preliminary suggestions of direction and topics for further reflection.

The focus of our interest in image studies is the recent work of Emmanuel Alloa. We imagine that some of aspects of his thinking, for instance, about how transparency theories and opacity theories of image developed historically in the European tradition, seem germane to our concerns with text-image relationships in Wen's album. But we imagine that instead of a relationship of inscription (something from Europe being written into the Chinese side), there might be relationships of exscription (something on the European side disjunctively and contrastively related to something on the Chinese side). Thus, what attracted us and encouraged us to persevere with Alloa's writings was the way he worked between Aristotle and Merleau-Ponty and especially the idea that images are opaque: images (and texts) are not transparent to an outside. When Alloa reviews a line of thinking about three kinds of seeing: seeing-as, seeing-in, and seeing-with,[25] we realize that it is too involved to be amenable to a summary account in a few sentences, but we were reminded of the way some Chinese historians of gardens had been stuck with simply 'seeing'. Having these terms before us reminded us that seeing is plural. It is in the multiple positioning of an ongoing engagement with situations and settings that we find the richness of garden experience. The eye is dislocated from its Cartesian position, becomes corporeal and dynamic. This seems to be what Wen's album is pointing to.

Ming dynasty connoisseurs likely engaged with these images by employing different layers of 'seeing through' their contemporary medium. This engagement extended beyond allusions, themes and artistic interests to encompass the interplay between text and image. Cross-reading of poems and paintings stimulates multiple layers of imagination, suggesting that the imagining of the garden develops differentially through the act of seeing with these distinct media.[26]

We can imagine that, for cultured connoisseurs, poems and paintings play a dual role in envisioning the garden. They independently articulate perceptions of the garden with distinct means and, in an intertextual resonance, evoke a novel understanding that is both unique and cohesive. Basically, readers of Wen's album have the option to read the garden through the lens of either poetry or painting, view a 'garden-image' through poetry, or encounter 'garden-poetry' through images, resulting in a complex interweaving of various experiential layers. Despite conventional decoding methods, such as thematic and formal analyses from an iconographic standpoint, the meaning of the images stays unresolved until textual identification. They accommodate diverse interpretations. When text operates as a mode of 'naming', the referential nature of poetry itself is indistinct, diverse, and open due to the extensive use of allusions and metaphors. Consequently, the semantic field of poetry effectively establishes a *milieu* for comprehending the paintings. Amidst the myriad references in poetry and the unbounded signifiers in paintings, numerous provisional referential connections emerge. These connections not only diverge from one another but also coexist, concreting as viewers perceive images with words and words reciprocate with images, echoing Merleau-Ponty's idea of *rhetorical chiasm*,[27] or, in Alloa's locution, a simultaneous co-implication.[28]

However, the way poetry and paintings unfold over time differs, indicating a lack of one-to-one correspondence between them. Furthermore, both lack symmetrical and simultaneous structures, revealing that the creation of the garden experience does not adhere to a specific predetermined form (*gestalt*) but happens gradually over time. François Cheng's examination of structural differences in regulated poetry highlights the intricate variations between couplets.[29] The first and fourth couplets follow a regular grammatical structure, guiding the progression of time. Conversely, the second and third couplets disrupt linear continuity with opposing phrases, creating a self-contained, lingering sense of space. The use of antithetical language connects allusions, imagination and reality, hovering between the lines and forming a relationship of intertextuality, allowing for the emergence of a dreamlike space. Spurred by the daydream, the 'closed time' of the first couplet transitions into the 'open time' of the fourth couplet, not just signalling the growth of understanding but also hinting at an irreversible passage of time. Cheng also observes that the space within the arrow of time (second and third couplets) transforms from static to dynamic[30] (Figure 3.4). This suggests that the fundamental structure of regulated poetry is not a symmetrical or fixed form but a dynamic process.

Here we can relate Alloa's analysis of chiasm in Charles Baudelaire's poetry '*À une passante*' ('To a Woman Passing By', 1855) to Chinese regulated poetry, emphasizing the perspectival nature of the latter: 'The chiasm is of a perspectival nature, and, as a result, it can always only be considered – as attested by its deployment over time – with reference to one of the terms in the relationship.'[31] This approach challenges the idea of a

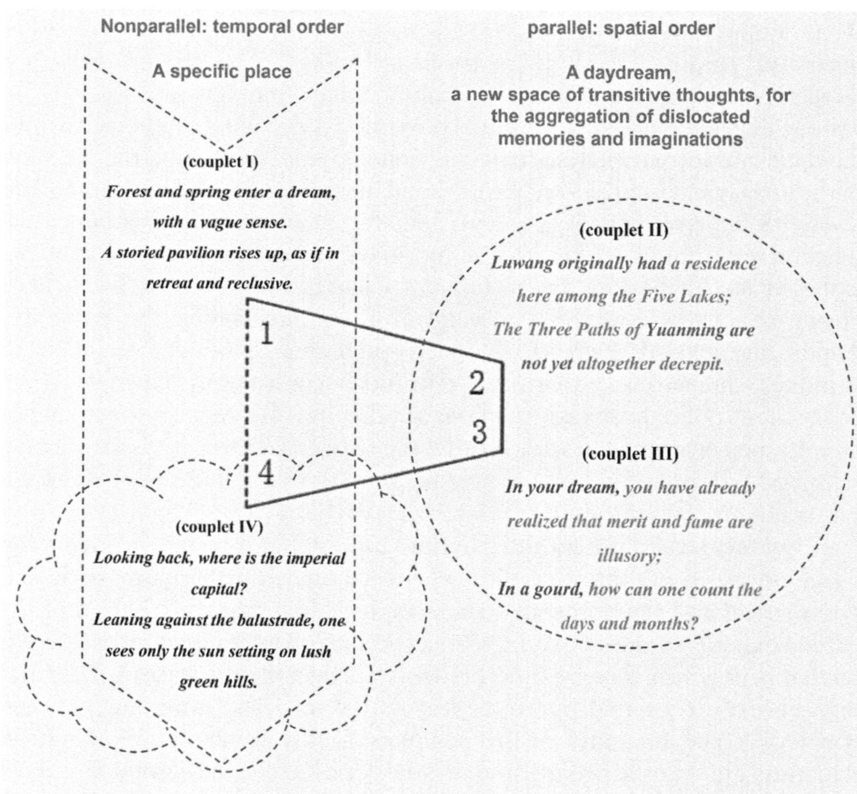

FIGURE 3.4 Diagram of 'The Storied Pavilion for Dreaming of Reclusion' following François Cheng's diagram (Hongde Wu).

symmetric and transparent reading, 'thus guaranteeing a divergence within inherence, a noncoincidence within simultaneity'.[32] Given the succinct style of the Chinese images, consisting of ambiguous landscape elements, their composition often aligns with themes of classical painting and should not be viewed as a literal illustration of the poetic verses.

Overall, the album does not offer a transparent understanding of historical reality, nor does it point towards a specific spatial form. Instead, garden experience is conveyed gradually through the implicit temporal progression found in both poems and paintings. Building on Merleau-Ponty's critique of transparency,[33] Alloa points out that cognition always takes place through an operative medium, whether it is the *Leib* (living body) or the *phenomenal milieu*, inherently involving resistance to the sensible world.[34] But what might be even more crucial is his distinction between focal seeing and lateral seeing in a milieu. He calls for a generative phenomenology 'that situates the origin of the potentiality of appearances in a perceptive laterality'.[35] The dream is lateral to the storied pavilion which, in turn is lateral to the sunset. A focus on a pagoda tree is displaced by the fragrance of its flowers spreading far and wide. The present scene is displaced by memories of official examinations and a former career-mindedness is displaced by relaxing in the cool of the night. The incidental sounds of rain and autumn, the light of dawn and the changing patterns of shade are lateral to the plantain and to architectural frames.

Conclusion

Through an analysis of three settings in the Garden of Artless Administration, it becomes evident that the interplay of 'repetition–difference' between poems and paintings cultivates a nuanced realm of perception and imagination. Consequently, the album transcends a mere representation of physical reality, aiming instead to evoke the latent potential inherent in embodied experiences within the garden. This allows readers not only to *see* the garden *in* the album but to *see* it *with* the album. Through immersive sensory experiences and the triggers of imagination and memory, readers gain a deeper understanding of the intertwined relationship between humans and gardens. This dynamic underscores the role of gardens as sanctuaries, facilitating the re-imagining of a meaningful life and tuning inner disposition and fostering virtuous actions. Furthermore, the social life revolving around gardens manifests a complex mechanism of representation, fostering shared interests among literati and preserving common sanctuaries within a fragmented world.

In sum, we encourage readers to go beyond superficial formal imaginings and myths of the formal restoration of gardens, and recommend the exploration of intricate relationships between material imagination and life in a garden retreat.

Acknowledgements

We would like to thank Dr Yue Zhuang for her kind invitation to present our work at the University of Exeter and for her patient help as we prepared our text for publication. We are grateful to Wang Hui and Lu Xiaoxuan for opportunities to present our work for discussion at Zhejiang University and The University of Hong Kong respectively. Students in Stanislaus Fung's course at Harvard University also responded to a presentation of our research with thoughtful and generous questions. Finally, we express our appreciation for the help that Jin Xuezhi, Ufuk Ersoy, Liu Runen, Xie Jie, Wen Zishen, Liu Shida and Pan Yiwei gave us in the final editing of our chapter.

Notes

1 Wang Chunyu, 'Lun Mingdai Jiangnan yuanlin' [On Gardens of the Jiangnan Area in the Ming Dynasty], *Zhongguo shi yanjiu*, 1987, no. 3.

2 See Pan Yiwei and Stanislaus Fung, 'Bibliography on the Garden of Artless Administration (Draft)', https://www.academia.edu/45168997/Bibliography_on_the_Garden_of_Artless_Administration_Draft_, accessed 1 April 2024.

3 Kate Kerby, *An Old Chinese Garden: A Three-Fold Masterpiece of Poetry, Calligraphy and Painting* (Shanghai: Chung Hwa Book Co., 1923). See also Wen Zhengming, *Wen Hengshan Zhuo Zheng Yuan shi hua ce* [Wen Hengshan's Album of Poetry and Paintings of the Garden of Artless Administration] (Shanghai: Zhonghua shuju, 1919).

4 See the note by Zhou Daozhen in Wen Zhengming, *Wen Zhengming ji* [Collected Works of Wen Zhengming], ed. Zhou Daozhen (Shanghai: Shanghai guji chubanshe, 1987), 1213.

5 Wei Jiazan, *Suzhou lidai yuanlin lu* [Record of Suzhou Gardens in Various Dynasties] (Beijing: Yanshan chubanshe, 1992), 109.

6 Gu Kai, *Mingdai Jiangnan yuanlin yanjiu* [Studies of Jiangnan Gardens in the Ming Period] (Nanjing: Dongnan daxue chubanshe, 2010), 73–7.

7 Craig Clunas, *Fruitful Sites: Garden Culture in Ming Dynasty China* (Durham: Duke University Press, 1996), 39. See also idem, 'Reading Wen Zhengming: Metaphor and Chinese painting', *Word & Image*, 25, no. 1 (2009): 96–102.

8 Runen Liu, 'L'allégorie de 'l'incapacité': La culture du jardin du Jiangnan dans la Chine impériale tardive, l'exemple du Zhuozhengyuan: Jardin de l'Activité politique d'un Incapable' [The Allegory of 'Incapacity': Garden Culture of Jiangnan in Late Imperial China, the Example of Zhuo Zheng Yuan: 'Garden of Artless Administration'] (PhD thesis, Université Paris-1 Panthéon-Sorbonne, 2011). See also, Mark Jackson, 'Landscape/Representation/Text: Craig Clunas's *Fruitful Sites*', *Studies in the History of Gardens & Designed Landscapes* 19, nos 3–4 (1996): 302–313.

9 Wei Xiuyu, *Guya kongjian: Wen Zhengming 'Zhuo zheng yuan san shi yi jing tu' yanjiu* [Wen Zhengming's 'Painted Album of Thirty-One Scenes of the Zhuo Zheng Yuan] (Beijing: Renmin chubanshe, 2017).

10 Andong Lu, 'Deciphering the Reclusive Landscape: A Study of Wen Zheng-Ming's 1533 Album of the Garden of the Unsuccessful Politician', *Studies in the History of Gardens & Designed Landscapes*, 31, no. 1 (March 2011): 40–59.

11 Wen Zishen and Stanislaus Fung, '*Yuanlin tiyan de zai lijie: Wen Hengshan Zhuo Zheng Yuan shihuace tandu*' [Rethinking Garden Experience: Preliminary Reading of *A Painted Album of the Garden of Artless Administration by Wen Zhengming*], *Zhongguo yuanlin*, 37, no. 4 (2021): 30–5.

12 Emmanuel Alloa, *Resistance of the Sensible World: An Introduction to Merleau-Ponty*, trans. Jane Marie Todd, Perspectives in Continental Philosophy (New York: Fordham University Press, 2017), x–xii.

13 François Cheng, *Chinese Poetic Writing*, expanded edition, trans. Donald A. Riggs and Jerome P. Seaton (Hong Kong: The Chinese University of Hong Kong Press, 2016), 76–7.

14 See Dalibor Vesely, 'The Hermeneutics of the Latent Word of Architecture', in idem, *The Latent World of Architecture: Selected Essays*, ed. Alexandra Stara and Peter Carl (London & New York: Routledge, 2023), 129–30.

15 Nathaniel Coleman introduced the idea of 'material imagination' as a counterpoint to 'formal imagination'. It emphasizes engaging with the tangible qualities and cultural meanings of architectural materials, prioritizing bodily experience over abstract concepts. Influenced by authors such as Bachelard, Eco and Semper, it underscores the poetic and symbolic significance of materials in shaping lived experiences. This concept challenges the notion of architecture solely as image-making or technical exercises, highlighting the emotional and practical value of built environments. See Nathaniel Coleman, *Materials and Meaning in Architecture: Essays on the Bodily Experience of Buildings* (London & New York: Bloomsbury Visual Arts, 2020), 3–5.

16 A *pingzuo* refers to the balcony that protrudes outward from the upper floor of pavilion-style buildings in traditional Chinese architecture. It is mainly composed of platforms, railings and brackets from the outside and usually indicates a mezzanine level inside. In her *Chinese Architecture: A History*, Nancy Steinhardt uses the transliteration 'pingzuo' and alternately describes it as a 'porch-like projection' or 'mezzanine level', depending on contextual nuances. See Nancy Steinhardt, *Chinese Architecture: A History* (Princeton, NJ: Princeton University Press, 2019), 49 and 164.

17 'Painting of Dwelling in a Storied Pavilion', or *lou ju tu*, is a recurring theme in the history of Chinese painting. It often portrays the serene lifestyle of reclining or residing within a pavilion perched atop a lofty structure, overlooking distant mountains and waters. This theme embodies the ideals of seclusion, introspection, and a harmonious communion with nature, evoking a profound sense of tranquillity and spiritual elevation. An early example can be found in the *Album Leaf of Spring View from a Storied Pavilion* (*Ceng lou chun tiao tu ye*) by the artist Yan Wengui (967–1044) of Northern Song Dynasty. (An impression of this painting can be obtained online at http://www.360doc.com/content/23/0825/03/42788182_1093775158.shtml,

accessed 7 August 2024.) Wen Zhengming himself has painted a work (see book cover image) entitled *Living Aloft* (*Lou ju tu*) for his friend Liu Lin in 1453. (Now held in the Metropolitan Museum of Art, New York.) In a colophon, he wrote: 'Mr. Liu of Nantan bid farewell to his official duties and plans to retire to a storied pavilion in order to make plain his lofty aspirations. Though the pavilion has yet to be constructed, I compose a poem and express its idea in a painting to herald it. In the future, the scroll shall adorn the right side of Mr. Liu's seat [in the completed building], enhancing the ideal of dwelling in a pavilion.' See Li Na, '*Wen Zhengming* Lou ju tu *kao: jianlun Mingdai wenren de shenghuo zhiqu*' [Study of Wen Zhengming's *Living Aloft*, With a discussion of aspirations and interests in the lives of Ming dynasty literati], *Jiangsu shehui kexue*, no. 6 (2015): 209–15.

18 William H. Nienhauser, *Tang Dynasty Tales: A Guided Reader* (Singapore and Hackensack, NJ.: World Scientific, 2010), 1:73–130.

19 See also Stanislaus Fung, 'Self, Scene and Action: The Final Chapter of *Yuan Ye*', in *Landscapes of Memory and Experience*, ed. Jan Birksted (London & New York: Spon Press, 2000), 131.

20 David L. Hall and Roger T. Ames, 'The Cosmological Setting of Chinese Gardens', *Studies in the History of Gardens & Designed Landscapes*, 18, no. 3 (July–September 1998): 177–83.

21 Nienhauser, *Tang Dynasty Tales*, 1:131–88.

22 Cheng, *Chinese Poetic Writing*, 76–7.

23 See, for example, Ji Yuan, Qin Renqiang and Huang Shun, '*Wenren shanshui yuanlin de guanzhao yu liumian: Jiyu* 'Zhuo Zheng Yuan sanshiyi jing tuce' *de quanjingshi fuyuan*' [Frontal and oblique views in literati gardens: Panoramic reconstruction based on *Thirty-one Scenes of the Garden of Artless Administration*], *Zhongguo yuanlin*, 32, no. 9 (2016): 87–93.

24 Wen and Fung, '*Yuanlin tiyan de zai lijie*.'

25 Emmanuel Alloa, *Looking Through Images: A Phenomenology of Visual Media*, trans Nils F. Schott (New York: Columbia University Press, 2021), 270–4. Emmanuel Alloa, 'Seeing-as, Seeing-in, Seeing-with: Looking Through Pictures', in *The Palgrave Handbook of Image Studies*, ed. Krešimir Purgar (Cham, Switzerland: Palgrave Macmillan, 2021), 483–99.

26 Ibid.

27 Maurice Merleau-Ponty, *The Visible and the Invisible, followed by Working Notes*, ed. Claude Lefort, trans. Alphonso Lingis (Evanston, IL: Northwestern University Press, 1968), 266.

28 Alloa, *Resistance of the Sensible World,* 72.

29 Cheng, *Chinese Poetic Writing*, 76–7.

30 Hongde Wu notes: The structure proposed by Cheng is aimed at regulated poetry (*lüshi*), which aligns with the analysis of the three cases discussed in the present essay. However, different techniques may be employed by the poet. For example, unlike the first two cases, which rely heavily on literary allusions, 'The Plantain Balustrade' opts for a direct description of specific objects. However, the 'asymmetry of visibility' between poetry and painting still propels

thoughts into a transitive, evolving process. The elements of the second couplet, 'It does not mind the high white wall, / And is well suited to the red balustrade that makes a zig-zag', are presented in both poetry and painting, establishing a setting (though the wall and balustrade evoke richer imagination in poetry than the limited visual expressions do). In contrast, the phenomenon described in the third couplet, 'The sound of winds from the north reaches one's pillows in autumn. / Green shafts of light break up the window lattices at dawn', extends beyond the painting, introducing a dynamic space only in the text. This then sets the stage for the cautionary advice about future actions in the final couplet. Thus, the subtleties of the 'here and now' milieu are gradually revealed through nuanced differences, which also align with the introduction of the three temporal scales in the initial couplet, each time leading to a finer granularity. This demonstrates that the stability of Cheng's structure does not contradict the rich techniques of regulated poetry. Compared to the heavy use of 'ready-made' literary references, grasping the instant differences in daily experiences is a more creative and nuanced approach, akin to the methods employed by poets such as Wang Wei and Bai Juyi. Elsewhere in Wen Zhengming's album there are other forms of poetry such as quatrains (*jueju*) and ancient-style poetry (*gushi*), which may not conform to Cheng's structure.

31 Alloa, *Resistance of the Sensible World*, 72.

32 Ibid.

33 Merleau-Ponty, *The Visible and the Invisible*, 114.

34 Alloa, *Resistance of the Sensible World*, 20–8.

35 Alloa, *Looking Through Images*, 211.

4

The gardens of Omar Khayyam

Poetic botany between Khurasan and Kathmandu

Iván Szántó

Gardens and poetry are microcosms in which the entrant may leave behind mundane concerns; they suspend external realities and offer an alternative. As for Persia, artificial landscapes and versified texts are two of its definitive cultural ingredients, the metaphorical quintessence of Persia often being imagined to lie in their conjunction.[1] They generated sensory crossovers already before the Muslim conquest of Iran in the seventh century BCE, although the exact time and conditions of this early coalescence are not entirely clear. While gardens (P. *bagh*) are documented since the establishment of the Achaemenid Empire (sixth century BCE), garden imagery in Persian poetry is connected to the rise of New Persian as the literary language of Islamic Iran from the eighth century CE onwards.[2] The earliest references date back to about 1000 CE, some decades before Omar Khayyam (1048–1131), one of the best-known figures of Persian poetry, would flourish.[3] Before dashing through a range of gardens which in one way or another were inspired by him, it is worth recalling that Khayyam's first recorded invocation of gardens – the foretelling of the location of his own tomb in Nishapur, Khurasan – was an astrological rather than a poetic feat; it preceded his death by twenty-four years and the first extant datable manuscripts containing verses from his *Ruba'iyat* by more than two centuries.[4] While his burial site still preserves some of its original aspects, the authenticity of much of Khayyam's poetry has been questioned, even aside from the *Rubáiyát*, Edward FitzGerald's (1809–83) Victorian transmutation. Chronological issues and authorship of the *Ruba'iyat*, however, are not tackled by this garden-oriented study. Instead, examples will be shown of how the synaesthesia of gardens and poems can create miniature Irans anytime and anywhere. This is why an essay about Persian *baghs* can easily divert from Persia and find another point of departure in, for instance, Kathmandu, Nepal. The present study starts out from there, flying from the farthest and most recent garden back to Khayyam's own burial ground like the nightingale of Persian poetry hops from one rose to another, and raises the question whether these are merely actualizations of a single garden: the garden of Omar Khayyam.

In the smoggy hustle and bustle of the Thamel district of Nepal's capital, still reeling from the 2015 earthquake, exhausted locals and visitors alike find a brief respite in the green oasis popularly known as the Garden of Dreams (Swapna Bagaicha). Still concealed from the city by its century-old walls – but no longer inaccessible for ordinary people – the garden forms an annexe to the Kaiser Mahal, a residence of Field Marshal Kaiser (Keshar) Shumsher Jung Bahadur Rana (1892–1964), the son of one of Nepal's hereditary prime ministers (Figure 4.1).[5]

Like the palace, the garden is neo-classical in style and seemingly devoid of anything typically Asian, let alone Nepali, apart from, of course, most of the greenery, as well as some faint resemblance to temple plans and building

FIGURE 4.1 Kaiser Mahal, central pavilion, Kathmandu, Nepal. Photograph: Author.

techniques. Due to decades of neglect in the later twentieth century, its current area is much smaller than its original size, and its layout differs from the original 1920 plan, at that time having comprised six pavilions to represent the six climes or six seasons of Nepal (hence its alternative name, 'garden of the six seasons').[6] In a quiet corner of this princely retreat, at a distance from the cascades, fountains and pergolas, on a wall of the porch of the central casino (the so-called Basanta Pavilion), there is a battered marble slab with an inscription in English, conveying a personal 'message' of the founder. The text comprises four separate quatrains from the Edward FitzGerald rendition of the *Rubáiyát* of Omar Khayyam (Figure 4.2).[7]

FIGURE 4.2 Marble slab with lines from *The Rubáiyát of Omar Khayyám* of Edward FitzGerald on the central pavilion of the Kaiser Mahal, Kathmandu, Nepal. Photograph: Author.

The slab is interlaced with engraved vines and tendrils which are said to camouflage cracks resulting from an earlier (1934) earthquake and probably inspired by the hand-painted and handwritten 1872 edition of the poem, by William Morris (1834–96) and Edward FitzGerald.[8] Juxtaposing the cyclical nature of time and the finiteness of life, this selection from the

Rubáiyát conjures up Khayyam/FitzGerald's whole work for those familiar with it and conceals the unwritten verses from the uninitiated. The unwritten lines matter, too, for Kaiser Rana might well have had a further quatrain with a more personal touch in mind when he created his hermitage and contemplated transience through Khayyam:

> I sometimes think that never blows so red
> The Rose as where some buried Caesar bled;
> That every Hyacinth the Garden wears
> Dropt in its Lap from some once lovely Head.[9]

It is noteworthy that the only non-Persian name in the *Rubáiyát* is that of Caesar's – the namesake of Kaiser Rana. It is so non-Persian a name that, in fact, it occurs only in FitzGerald's 'translation', and nowhere in the Persian original. In the blood-soaked era of the hereditary Rana prime ministers, such a reference was better to have remained unuttered; still, it may well have resonated with erudite visitors. Although Khayyam's Persian text does not mention Caesar here or elsewhere in the *Ruba'iyat*, the name itself was familiar in the Arabo-Persian world, just like in its Hindustani extension, as reflected by the Persianate name of Kaiser Shumsher Rana. Whatever we think about this coincidence, there was no chance that this quatrain could be displayed: not only the explicit reference to Caesar's blood was rather ill-suited for a member of the Rana dynasty, but the mention of a burial made it a taboo for even the most outspokenly Westernized Hindu aristocrat. Thus, this quatrain was (and is) there only for those who knew the entire *Rubáiyát*, and shared its supposed agnosticism.

With this garden, Kaiser Rana fashioned himself as an Omar Khayyam of his times: a Nepali sage and a European gentleman at the same time; 'he was every bit the image of a slightly Oriental Voltaire, but he possessed the shrewdness of a Talleyrand allied to the intelligence and culture of a French academician', as one of his guests noted.[10] Coming to power as a result of the Kot massacre in 1846, the Rana regime (1846–1951), with British help, sidelined the Nepali ruling house which nominally kept its throne. Its British support is indicated by the fact that, as soon as he consolidated his power, the founder of the regime, Prime Minister Jung Bahadur Rana (in office 1846–56 and 1857–77) became the first South Asian leader to pay an official state visit to London in 1850, followed by a detour to Paris in the same year.

Kaiser Rana may also have been familiar with the writings of Eben Francis Thompson (1859–1939), the founder (in 1900) and president of the Khayyam Society of America, who, in the preamble of *The Rose Garden of Omar Khayyam*, his own version of the *Ruba'iyat* 'founded on the Persian' and written ten years before the establishment of the Kathmandu grove, describes how the retired Khayyam, 'King of the Wise', lived in the seclusion of a garden sanctuary filled with science and overgrown with vines of rose and grape.[11] It was in this *bagh*, Thompson imagined, that Khayyam received his most devout admirers and disciples. Although Kaiser Rana was still a young man in 1920, he fashioned his garden as the imaginary home of a venerable Omar Khayyam; the Kaiser Mahal still comprises the Kaiser Library north of the garden, which at the time of its establishment was the most comprehensive library in Nepal, full of rare old manuscripts and modern scholarship, and strictly off-limits for everyone except a selected few, but, as stipulated in the will of the founder, with a prospect of general accessibility after the passing away of the founder.[12] Scholarship alone would not have made Kaiser Rana worthy of a Khayyam, but 'the Field Marshal's passion for knowledge was equalled only by his love of refinement. A gourmet, he had one of the best wine cellars in Asia. His taste for beautiful women and fine food was as famous as his political and diplomatic ability'[13] – and his penchant for breaking taboos. Immersing themselves in the enjoyment of food and drink, while cherishing the illusion of standing at the pinnacle of the intellectual elite, was a characteristic attitude of early twentieth-century Khayyam Club members.[14]

As a man of learning, Kaiser Rana may also have been familiar with and inspired by other Persian classics. Indeed, the garden's original six pavilions, corresponding to the six seasons of Nepal, might allude to the *Seven Princesses* (*Haft paykar*) by Nizami Ganjavi (*c.* 1141–1209), in which King Bahram – mentioned also by FitzGerald along with Caesar – builds seven distinctly coloured pavilions in his palace gardens for his seven princesses.[15] Each princess was sent by one of the seven rulers of the world, including one by the Caesar of Rome. The king visited one princess every day to listen to the tales they told, while enjoying wine. He visited the princess of Rome on Sunday in the yellow pavilion, reflecting the colour of the Sun, the dominant planet of that day.[16]

The title of Nizami's poem refers to seven portraits: these depicted Bahram Gur's future wives in an unlocked secret chamber and the paintings were so lifelike that the king fell in love with the seven princesses all at once. While this was a literary trope, such imagery could find its way to real life as well in medieval Islamic art. In the Early Islamic bathhouse of Qusayr ʿAmra (Jordan), once surrounded by an orchard in the middle of the desert, we see not only the depiction of a bathing lady who has been interpreted as a captive Persian princess, but also six defeated rulers, among them the emperor (*qaysar*) of Rome.[17] Indeed, the Umayyad caliphs (r. 661–750) liked to regard themselves as the inheritors of the Roman emperors to the extent that their adversaries expressly condemned them for their Caesarism (*qaysariyya*).[18] This contributed in Arabo-Persian poetry to the appearance of the fatalistic image of Caesars buried under the rubble of their bygone palaces. It is unclear whether Kaiser Rana knew about Qusayr ʿAmra and whether his library did possess books about it, but Qusayr ʿAmra was one of the most explosive subjects for debate in international art historiography during the 1910s and 1920s, making the secret pleasure-ground of the caliphs much more famous than the Umayyads ever wanted it to be. In fact,

Alois Musil (1868–1944), the Austro-Moravian discoverer of the site (1898), who was never able to return to Qusayr ʿAmra after the fall of the Habsburg Empire, took refuge in fiction, as he wrote a novel about a medieval Bohemian noblewoman's love affair with the Lord of ʿAmra.[19]

Were it not for the poem and the pavilions (in Iran, gardens and pavilions being inseparable), the Kaiser Garden would lose much of its metaphoric dimension: while in Persia the impact and allusive potential of gardens derive from their struggle with their arid surroundings, in Nepal mountains and monsoon create a biodiversity which makes the time-honoured art of Persian horticulture irrelevant. In his private Oriental paradise, Kaiser Rana basked in the glory of the British Empire and shunned all things related to traditional Hindu society, except for its rigid social structure. He had no intention of creating a particularly Persian microcosm in Nepal.[20] Yet, for others, this was a true perspective. Just a stone's throw from the Kaiser Mahal, one can find the earliest mosque in the Kathmandu Valley, the Kashmiri Takiya Jami Masjid (Figure 4.3).

As its name implies, the mosque was founded by Kashmiri settlers; in fact, written sources show that the first Muslims in Kantipur (now Kathmandu) were Kashmiri wool merchants around 1500, who used Nepal as a stopover between their homeland and Lhasa.[21] Soon thereafter they would receive various positions at the court of King Ratna Malla (r. 1482–1520). In

FIGURE 4.3 Funerary enclosure of Shah Ghiyath al-Din Kashmiri, Kathmandu, Nepal. Photograph: Author.

particular, this king fell under the spell of a Kashmiri Sufi mystic, Shah Miskin Baba (fl. early sixteenth century), who could secure precious land near the royal palace where he established the mosque in 1524 and where his and his nephew and successor's (Shah Ghiyath al-Din Kashmiri) shrines would be erected: these are called Rawza ('garden' in Arabic/Persian) and are still surrounded by lush greenery and replete with Persian poetic inscriptions, bearing testimony to the strength of cultural Iranism in Nepal.

Shah Miskin Baba's project stopped short of a widespread Islamization of the Kathmandu Valley, and there is no organic chronological development between the Kashmiri shrine and the Kaiser Mahal. But in the western Himalayas, in the Kashmir Valley proper, such organic developments did take place more than a century earlier, outlining the eastward spread of Persianate Islam from Central Asia.[22] Tradition attributes this achievement to another Sufi master, Mir Sayyid ʿAli Hamadani (1314–84), a Kubrawi mystic who, during his lifelong peregrinations, traversed Western, Central and Southern Asia between his native Hamadan in Persia and Kashmir, the seat of his lodge (*khanaqa*), using this route to encompass these outlying lands in a common spiritual sphere.[23] Owing to his presence on this pre-existing commercial route between Persia and the Himalaya, the road also offered access to the supernatural, and Mir Sayyid ʿAli Hamadani came to be remembered and venerated chiefly for establishing this sacred link. In the Persian words of the modern poet Muhammad Iqbal (1877–1938):

> Guide he of that emerald land [Kashmir],
> counsellor, of prince and dervish and sultan;
> a king ocean-munificent, to that vale
> he gave science, crafts, education, religion.
> That man created a miniature Iran
> with rare and heart-ravishing arts.[24]

For Iqbal, these contributions are inseparable from Islam, or more precisely, from its Persianate Sufi version. Yet science, crafts, education and religion had many centuries of history in Kashmir before the arrival of Islam, and more earthbound commenters may consider Mir Sayyid ʿAli Hamadani a political refugee from Central Asia, most likely exiled by Timur (r. 1380–1405), rather than the 'civiliser' of Kashmir.[25] Still, in the Persianate world, civilization is envisaged as the coalescence of Persian gardens and Persian poetry and the best epitome of this is a Persian couplet which has almost become the motto of Kashmir: 'If there is a Paradise on Earth, it is here, it is here, it is here'.[26]

This oft-quoted distich was known already in the seventeenth century to have been composed by India's most famous Persian poet, Amir Khusraw Dihlawi (1252–1325) and inscribed in the so-called Black Pavilion of the Shalimar Garden in Kashmir.[27] The problems with this example of historical memory are that Amir Khusraw's collected works do not contain this couplet; the poet never visited Kashmir, which was not even Muslim during his time; and the poem is not written anywhere in the Shalimar Garden, which would be established in the seventeenth century.[28] Instead, the text is visible in the Divan-i Khass of the Red Fort in Delhi.[29] But gardens and poetry create associations, arising from the rhyming of natural and poetic imagery, and resulting in a would-be reality in which the separate realities of a perfect line in Persian, the greatest Hindustani master of Persian verse, and the most beautiful spot in Kashmir become one.

Flying to the next valley from Kashmir to the west, approaching the land of Omar Khayyam, a layover is required in the Kabul Valley, which until modern times was dominated by a garden, closely associated with political conquest and Persian poetry (Figure 4.4).

The Bagh-i Babur is at once an imperial monument, the scene of major social events, a memorial of a quest for solitude and earthly permanence, and, ultimately, a place of burial.[30] It is the creation of the erudite founder of the Mughal Empire, Zahir al-Din Babur (1483–1530, r. 1526–30), on the only foothold he could cling to after he had lost his inheritance of the Central Asian garden cities of Samarqand and Herat and before he would conquer Delhi, at a moment when he was yet to become Padshah. As a Timurid nomad, his Kabuli garden served as a campsite rather than a palace ground and this notion of transitoriness immediately implies Khayyam-like

FIGURE 4.4 *The tomb of Emperor Baber*. Lithograph by Louis and Charles Haghe after drawing by James Atkinson. Atkinson, James. *Sketches in Afghaunistan*. London: Henry Graves & Company,1842, plate 24. Library of Congress Prints and Photographs Division Washington, D.C. Reproduction Number: LC-DIG-ppmsca-41535

overtones. Babur himself, in his memoirs, the *Baburnama*, records how he employed his garden as the stage of drinking ceremonies in order to strengthen the loyalty of those on whom he counted for his forthcoming campaigns.[31] After his victory at Panipat (1526) over the Hindus, whom he called unbeliever (*kafir*), he held a Ramadan ceremony in another newly established garden at Fathpur Sikri.[32] Here is another, more radical, instance when the garden has a specifically Perso-Muslim decorum in a Hindu environment. At Panipat, the site of his major victory, he planted a garden in an area which is still called Kabuli Bagh, implying a metaphorical unity between gardening and conquest.[33] His lost (but in 2008, rebuilt) tomb in Kabul was an open-air resting place far in scale (and location) from the grand mausolea of his imperial descendants. His remains were hastily reburied there before 1544 on order of his fleeing son, the Emperor Humayun (r. 1530–44, 1555–6), who left Agra (where Babur died and was buried in 1530) for Qandahar.[34] The tomb was a modest starting point for the later Mughal mausolea: an enclosure without a roof for an exile even in death. Later Mughal rulers would upgrade it; the first major alteration of Babur's garden and tomb occurred in 1675 when Shah Jahan (r. 1628–58) built a white marble mosque opposite the tomb to mark his conquest of Balkh.[35]

Omar Khayyam's trail is picked up in Balkh, as it was in this city that one of the very few recorded encounters with him occurred. Accordingly, Nizami ʿAruzi Samarqandi (fl. *c.* 1110–60) recalls in his *Chahar Maqala* ('Four Discourses') how he was overcome with a mixture of emotional turmoil and astonishment when in 1135/6 he found the tomb of his one-time master in astrology in the city of Nishapur, just as the astrologer had foretold it twenty-four years before. 'My grave will be in a spot where the trees will shed their blossoms on me twice a year', Nizami quotes Khayyam, before continuing:

> when I arrived at Nishapur in the year AH 530, it being then four years since that great man had veiled his countenance in the dust, and this nether world had been bereaved of him, I went to visit his grave on the eve of a Friday . . ., taking with me one to point out to me his tomb. So he brought me out to the Hira Cemetery; I turned to the left, and found his tomb situated at the foot of a garden-wall, over which pear-trees and peach-trees thrust their heads, and on his grave had fallen so many flower leaves that his dust was hidden beneath the flowers. Then I remembered that saying which I had heard from him in the city of Balkh, and I fell to weeping.[36]

While such experiences of premonition were rare in the time of Nizami, they would become more common with the spread of Sufism in the thirteenth and fourteenth centuries. By this time scientific prognosis was superseded by mystical time travel, as shown by an inscribed plaque from a garden in Kashan (dated 1312 CE); the plaque records a dream in which ʿAli himself (*c.* 600–61), the first imam of Shia Islam and the fourth caliph of Sunni Islam, orders the construction of his own memorial according to specific dimensions and in a strictly prescribed location in the otherwise unknown Mihriqabad garden, at a stopover which took place during a night journey from his burial site in Iraq to a yet-to-be-converted India. This event happened 650 years *after* the death of ʿAli.[37]

Omar Khayyam the scholar had, thus, already set the stage for an idiom before the poet Omar Khayyam would even be born; given that in his own time (and until the nineteenth century) he was celebrated chiefly as an astrologer, such a feat would not have been considered to have been beyond his professional competence. In medieval Islamic scientific thought, astrology was regarded as the ability to acquire the power of forewarning, to choose auspicious moments and favourable localities. It was not until the growing international fame of Edward FitzGerald's *Rubáiyát* that Khayyam metamorphosed into one of the greatest Persian poets.[38] Unsurprisingly, the tomb of a thousand-year-old astrologer/mathematician, even the greatest one, was not well maintained and venerated throughout the centuries and its mere survival in the otherwise harshly decimated Nishapur into the twentieth century was almost a miracle. It did survive, though, and the ultimate destruction and replacement of the grave was the result of Khayyam's Western-inspired rediscovery as a poet.[39] In 1892 George Nathaniel Curzon (1859–1925) found it standing 'in a neglected garden formerly containing flower-beds and rivulets of water, but now a waste of weeds'.[40] A string of European and American 'pilgrims' came in the early twentieth century to pay respect and all lamented the miserable condition of the tomb, the oblivion of Iranians, and, to quote one of the visitors, that 'no roses fell on his tomb as he [= Khayyam] had hoped'.[41] The sense of guilt of the Iranian heritage authority was surely aggravated by news that international Khayyam enthusiasts created a plethora of lavish albums, gentlemen's clubs, and other memorials in his honour across the world: the Garden of Dreams of Kaiser Rana was just one of them. Already in 1884, William Simpson (1823–99), an admirer of the *Rubáiyát*, and an artist working for the *Illustrated London News* and reporting about the work of the Afghan Boundary Commission on the Russian-Afghan border, took the opportunity to visit Omar Khayyam's grave. He was able to collect 'some hips and sent them to Bernard Quaritch (1819–1899), who passed them on to Kew Gardens, where the Keeper raised new plants from the seeds'.[42] In 1893, a year after the foundation of the London Omar Khayyam Club, Simpson, Quaritch and other members planted a rose from Kew on FitzGerald's grave in the churchyard at Boulge (a few miles from Woodbridge, Suffolk).

More than anything, it was a feeling of shame in front of these European visitors that compelled the leaders of the nascent heritage organization of the modernist early Pahlavi state to step into action in 1931, when they commissioned the German-trained architect Karim Tahirzada Bihzad (1888–1963) to create a new, Europeanizing, memorial worthy of the 'poet' Khayyam.[43] The original one, which was built according to Khayyam's own intentions, maintained for nearly a millennium, and made possible its

FIGURE 4.5 Funerary enclosure of Omar Khayyam. Nishapur (Khurasan-i Razavi Province, Iran). Photograph: Eszter Csontos.

identification, was demolished without proper documentation in 1934.[44] The monument and the *imamzada* which originally included it was now made accessible via a 150-meter treelined asphalt road. Ironically, Bihzad's monument was said to have been so carelessly constructed that merely thirty years later, in 1963, it would also be dismantled to give place to an Islamo-modernist structure by Houshang Seyhoun (1920–2014), which still stands today.[45] This domical structure expressly aims to amalgamate Khayyam's scientific and poetic achievements and fulfil his wish, as its pierced openings allow the north wind to blow through it and bring blossoms (Figure 4.5).

Determining whether Omar Khayyam could have foreseen this transformation, including that of his posthumous fame, and the gardens he inspired awaits an ivory tower scholar untouched by the metaphysics of the botanical microcosm.

Notes

1 Elisabeth B. MacDougall and Richard Ettinghausen, eds., *The Islamic Garden* (Washington, DC: Dumbarton Oaks Trustees for Harvard University, 1976); Sheila S. Blair and Jonathan M. Bloom, eds., *Images of Paradise in Islamic Art* (Hanover, NH: Trustees of Dartmouth College, 1991); Attilio Petruccioli, ed., *Gardens in the Time of the Great Muslim Empires: Theory and Design* (Leiden, New York: Brill, 1997).

2 David Stronach, 'Parterres and Stone Watercourses at Pasargadae: Notes on the Achaemenid Contribution to Garden Design', *The Journal of Garden History*, 14 (1994): 37–41.

3 Charles-Henri de Fouchécour, *La description de la nature dans la poésie lyrique persane du XIe siècle: Inventaire et analyse des thèmes: Travaux de l'Institut d'Études Iraniennes de l'Université de Paris, 4* (Paris: Librarie C. Klincksieck, 1969), 34–8; Christine van Ruymbeke, *Science and Poetry in Medieval Persia: The Botany of Nizami's* Khamsa. (Cambridge: Cambridge University Press, 2007).

4 For a summary of Khayyam studies, see Ḥosayn Maʿṣumi Hamadāni et al., *Encyclopaedia Iranica Online* (New York: Columbia University Press, 2020), s.v. 'KHAYYAM, OMAR', https://referenceworks.brill.com/display/entries/EIRO/COM-12113.xml (accessed 5 January 2024); for FitzGerald, see William H. Martin and Sandra Mason, eds., *Edward FitzGerald's Rubáiyát of Omar Khayyám: A Famous Poem and its Influence* (London and New York: Anthem Press, 2011); for a particularly accurate, though scarcely known, overview of Khayyam problems, see Bertalan Csillik, *Omar Khajjám Rubáijátjának kisebb kéziratai a párisi Bibliothèque Nationale-ban. A Suppl. Pers. 1777., 826., 745., 793., 1481., 1425., 1817., 1317., 1458. számú kéziratok eredeti szövegei* (Szeged: Szeged Városi Nyomda és Könyvkiadó, 1933).

5 The site is treated in detail by the autobiography of its restorer, Götz Hagmüller. See Götz Hagmüller, *A Picture Book of my Life* (Kathmandu: Vajra Publications, 2017), vol. 2, 1–50. I am indebted to Ms Ludmilla Hungerhuber, wife of Hagmüller, for sharing her recollections about the restoration of the garden between 1998 and 2007.

6 Hagmüller, *A Picture Book*, vol. 2, 5–9.

7 Edward FitzGerald, *Rubáiyát of Omar Khayyám, the Astronomer-Poet of Persia* (London: Bernard Quaritch, 1859), nos. XXXVIII, LXXII–IV; Hagmüller, *A Picture Book*, vol. 2, 11.

8 William H. Martin and Sandra Mason, *Encyclopaedia Iranica Online* (New York: Columbia University Press, 2009), s.v. 'KHAYYAM, OMAR vi. Illustrations of English Translations of the Rubaiyat', http://www.iranicaonline.org/articles/khayyam-omar-illustrations-of-english-translations (accessed 5 January 2024); Michaela Braesel, 'William Morris, Edward Burne-Jones and *The Rubaiyat of Omar Khayyam*', *Apollo*, 159, no. 2 (February 2004): 47–56.

9 FitzGerald, *Rubáiyát*, no. XVIII.

10 French anthropologist Michel Peissel's testimony about the elderly Kaiser Rana, quoted in Stefanie Lotter, 'Distinctly Different Everywhere: Politics of Appearance Amongst Rana Elites Inside and Outside Nepal', *Comparative Sociology*, 10 (2011): 523.

11 Eben Francis Thompson, *The Rose Garden of Omar Khayyam* (Worcester, MA: private publication, 1910).

12 Dragomir Dimitrov and Kashinath Tamot, 'Kaiser Shamsher, his Library and his Manuscript Collection', *Newsletter of the NGMCP* [Nepal-German Manuscript Cataloguing Project] no. 3 (January–February 2007): 26–36.

Home of more than 80,000 volumes and 600 old manuscripts, the library was heavily damaged by the 2015 earthquake and is currently operating in a modest temporary facility.

13 Michel Peissel, quoted in Lotter, 'Distinctly Different', 523.

14 Starting with the foundation of the Omar Khayyam Club of London in 1892, similar clubs sprang up in many countries during the early twentieth century, as a mark of the worldwide devotion to the *Rubáiyát* and its cult of sensual pleasures. Michelle Kaiserlian, 'The Imagined Elites of the Omar Khayyám Club', in *Edward FitzGerald's Rubáiyát of Omar Khayyám: Popularity and Neglect*, ed. Adrian Poole, Christine van Ruymbeke, William H. Martin and Sandra Mason (London and New York: Anthem Press, 2011), 147–9.

15 Julie Scott Meisami, 'Allegorical Gardens in the Persian Poetic Tradition: Nezami, Rumi, Hafez', *International Journal of Middle Eastern Studies*, 17 (1985): 229–60; Julie Scott Meisami, 'The Body as Garden: Nature and Sexuality in Persian Poetry', *Edebiyat*, 6 (1995): 245–74.

16 Paola Orsatti, 'The Order of Climes in Nezāmi's *Haft peykar*', in *Zur lichten Heimat. Studien zu Manichäismus, Iranistik und Zentralasienkunde im Gedenken an Werner Sundermann*, ed. Team 'Turfanforschung' (Wiesbaden: Harrassowitz, 2017), 483–90.

17 Qusayr 'Amra was built for the future caliph al-Walid II in the second quarter of the eighth century. Garth Fowden, *Quṣayr 'Amra: Art and the Umayyad Elite in Late Antique Syria* (Berkeley: University of California Press, 2004), 197–226.

18 Fowden, *Quṣayr 'Amra*, 236, n. 27.

19 Alois Musil, *Pán Amry* (Prague: Vyšehrad, 1948); Martina Veselá, 'Alois Musil: After Arabia and Vienna', in *Alois Musil: Interdisziplinäre Perspektiven auf eine Vielschichtige Persönlichkeit: Investigatio Orientis* 7, ed. Benedikt J. Collinet, Ludger Hiepel and Martina Veselá (Münster: Zaphon, 2021), 111–52.

20 Even so, by the fourteenth century, quatrains later to be attributed to Omar Khayyam would reach as far as China, judging from an inscription from Fujian province, Southeast China. Bernard O'Kane, *The Appearance of Persian on Islamic Art* (New York: Persian Heritage Foundation, 2009), 119.

21 Megan Adamson Sijapati, *Islamic Revival in Nepal: Religion and a New Nation* (London and New York: Routledge, 2011), 20–1.

22 For the expansion of Persian, see Nile Green, ed., *The Persianate World: The Frontiers of a Eurasian Lingua Franca* (Los Angeles: University of California Press, 2019).

23 Devin DeWeese, 'The Eclipse of the Kubravīyah in Central Asia', *Iranian Studies*, 21, no. 1–2 (1988): 51–4.

24 Sir Muhammad Iqbal, *Javid-Nama*, trans. Arthur J. Arberry (London: George Allen and Unvin, 1966), 116–17.

25 DeWeese, 'The Eclipse', 53–4.

26 Ebba Koch, 'Mughal Palace Gardens from Babur to Shah Jahan (1526–1648)', *Muqarnas*, 14 (1997): 160 n. 4.; Annemarie Schimmel, 'The Celestial Garden in

Islam', in *The Islamic Garden*, ed. Elisabeth B. MacDougall and Richard Ettinghausen (Washington, DC: Dumbarton Oaks Trustees for Harvard University, 1976), 20.

27 The history and layout of the garden are summarized in Attilio Petruccioli, 'Gardens and Religious Topography in Kashmir', *Environmental Design: Journal of the Islamic Environmental Design Research Centre (Rome)*, 1–2 (1991): 71–3.

28 The verse is attributable to Mirza Muhammad Tahir Khan Ashna (1627–70) whose description of Kashmir in his abridged Persian *Shahjahannama* is replete with garden imagery and poetic allusions; see Mirza Muhammad Tahir Khan Ashna (Inayat Khan), *Mulakkhas-i Shahjahannama*, ed. Jameel-ur-Rehman (New Delhi: Centre for Persian Research, 1388 [2010]), 186–90.

29 Koch, 'Mughal Palace Gardens', 158.

30 Salome Zajadacz-Hastenrath, 'A Note on Babur's Lost Funerary Enclosure at Kabul', *Muqarnas*, 14 (1997): 135–42.

31 Analysed in Catherine B. Asher, 'Babur and the Timurid Char Bagh: Use and Meaning', *Environmental Design: Journal of the Islamic Environmental Design Research Centre (Rome)*, 1–2 (1991): 50–1.

32 Asher, 'Babur', 47; Koch, 'Mughal Palace Gardens', 143.

33 Asher, 'Babur', 51–3.

34 Zajadacz-Hastenrath, 'A Note on Babur', 137.

35 For the migration of Timurid garden engineering techniques and professionals to Mughal India, see Maria E. Subtelny, 'Agriculture and the Timurid Chaharbagh: The Evidence from a Medieval Persian Agricultural Manual', in *Gardens in the Time of the Great Muslim Empires: Theory and Design*, ed. Attilio Petruccioli (Leiden and New York: Brill, 1997), 110–28.

36 Edward G. Browne and Mírzá Muḥammad of Kazwín, eds., *Revised Translation of the* Chahár Maqála *of Niẓámí-i-ʿArúḍí of Samarqand* (London: Luzac, 1921), 71–2.

37 The ceramic plaque is preserved in the Musée National de Céramique, in Sèvres; Chahryar Adle, 'Un disque de fondation en céramique (Kāšān, 711/1312)', *Journal Asiatique* 260 (1972): 280–5; Abdollah Ghouchani, *Ashʿar-i farsi-yi kashi-hayi Takht-i Sulayman*: *Persian Poetry on the Tiles of Takht-i Sulaymān* (Tehran: Iran University Press, 1371 [1992]), 12–15.

38 Jos Biegstraaten, *Encyclopædia Iranica Online* (New York: Columbia University Press, 2008), s.v. 'KHAYYAM, OMAR xiv. Impact on Literature and Society in the West', http://www.iranicaonline.org/articles/khayyam-omar-impact-west (accessed 5 January 2024); Adrian Poole, Christine van Ruymbeke, William H. Martin and Sandra Mason, eds., *Edward FitzGerald's Rubáiyát of Omar Khayyám: Popularity and Neglect* (London and New York: Anthem Press, 2011).

39 For a thorough survey of the modern history of the Khayyam memorial in Nishapur, see Talinn Grigor, *Building Iran: Modernism, Architecture, and National Heritage under the Pahlavi Monarchs* (Pittsburgh: Periscope Publishing, 2009), 146–63.

40 George Nathaniel Curzon, *Persia and the Persian Question* (London: Longmans, Green & Co., 1892), 263–4.

41 As referred to in Grigor, *Building*, 151. Others, however, at least in the 1880s, were delighted to find the roses.

42 Kaiserlian, 'The Imagined Elites', 151.

43 Grigor, *Building*, 154–7.

44 Ibid., 150–4; Shahryar Khanizad, *Hushang Sayhun* [Houshang Seyhoun]*: Mi'mar, Hunarmand, Naqqash* (Tehran: Hunar-i mi'mari, 1394 [2016]), 208–33.

45 Grigor, *Building*, 157–65. Karim Tahirzada Bihzad's work also survives on Khayyam Square in central Nishapur.

5

Finding solace in garden memories

Osvald Sirén's life during the Second World War

Minna Törmä

In a letter to Florence Ayscough (1878–1942) in 1930, Osvald Sirén (1879–1966) lamented his return to Sweden and to his day job as a curator in the Nationalmuseum in Stockholm after an extensive stay in East Asia 1929–30. He abhorred the (administrative) routines of museum work and was at times frustrated because of the resistance to his efforts to increase the presence of Chinese objects, particularly paintings, in the museum's collections. Nevertheless, he ended his letter on a more optimistic note, considering that he had during his time in China and elsewhere accumulated 'plenty of beautiful memories and interesting materials' and that to help to ease his situation he should attune himself 'to the spirit of T'ang [Tang] poetry and Sung [Song] painting'.[1] This quote displays one example of how Sirén created imagined space by returning to the world of memories and also of poetry and painting and using them as sites of escape and refuge.

The focus of this chapter is to explore the life of Sirén during the Second World War, when he lived isolated in his home on Lidingö, an island outside Stockholm.[2] When the war began, he was nearing retirement, which came in 1944. The narrative considers, through the example of Sirén's life in those years, how in times of conflict gardening and memories of landscape experiences can serve as a refuge. We might take our cue here from the often quoted first lines of Vita Sackville-West's (1892–1962) poem *The Garden*, written on and off during the war and published in 1946:

Small pleasures must correct great tragedies,
Therefore of gardens in the midst of war
I boldly tell . . .[3]

However, before developing the wartime phase of his life, it will be useful to take a closer look at the quote cited above in order to get a sense of how Sirén generally addressed less than genial states of affairs and escaped to an imagined space. During the decade that followed his return from China in 1930, Sirén increasingly devoted his energies to the study of Chinese painting, while in the previous decade, when for the first time he wholeheartedly embraced Chinese culture and travelled widely around the country, his publications had featured sculpture and architecture.[4] Sirén recognized that the understanding and the appreciation of painting required knowledge of poetry. However, he was by no means a stranger to poetry, as in his earlier years he had published two collections of his own poems: *Accord* (1902) and *Stilla stunder med naturen* (Quiet moments in nature; 1913).

Sirén did not write about his views on Tang poetry in detail, though his appreciation of it is implicit in his published research on Chinese painting. Glancing through *A History of Early Chinese Painting*, vol. I (1933), one of the sub-chapters (under the chapter on the Middle T'ang [Tang] Period) is titled 'The Golden Age of Art and Poetry'. In this short chapter, he employs the personalities of Du Fu (712–70) and Li Bai (701–62) to discuss the milieu of the reign of emperor Minghuang (Xuanzong, r. 712–56) and mentions Han Yu (768–824) as the greatest prose writer.[5] He picks up a few quotes from the lines of the poetry of Du Fu and Li Bai in order to create a nostalgic air of the period in these pages, emphasizing the shift from the 'golden years' of art and culture to a disaster (a conventional way to view the reign of Minghuang at that time). What else, in addition to nostalgia, Sirén read in their poems is difficult to judge based on the brief historical contextualization.

What about the spirit of Song painting then? When discussing Five Dynasties (907–60) and early Northern Song (960–1127) painters, Sirén emphasized that '[t]heir ideal was rather to create like nature herself, to visualize their ideas in shapes of mountains, waters, and trees, to make the pictures unfold or grow, as do things in nature, according to a certain plan of vision'.[6] But once he tackles their paintings (attributed paintings for most part), the analysis becomes more focused on formal properties. His enthusiasm increases when he reaches Southern Song painters; in that context Ma Yuan (active before 1189–after 1225) seems to reign supreme, at least for Sirén: 'There may have been greater painters in China, but no one who with a few strokes of the brush transformed more completely shapes of nature to symbols of unseen reality'.[7] Painters, like poets, draw their ideas from their surroundings and experiences transporting the viewer to imaginary worlds they have created. These imagined spaces – poems, paintings or gardens combined – were places where Sirén sought refuge.

Creating a garden

Sirén's souvenirs from sojourns in East Asia were visible inside his house, which had been enlarged and renovated while he himself was travelling there in 1929–30. The new additions were, among other details, a moon door connecting two sections of the living room and panels depicting phoenixes decorating the ceiling.[8] But he did not limit himself by paying attention only to the interiors when creating an appropriate environment for work, and souvenirs spilled beyond the walls of the house. Sirén had created his garden partly according to the ideas he had learnt on his visits to landscapes and gardens in East Asia during the early part of the twentieth century. Though Sirén's house still exists, its garden has been destroyed and modern buildings now cover part of the originally quite extensive plot.

The garden had been created before the start of the Second World War, but it gained vital importance during the war years. Since travel during the war was restricted – practically impossible – Sirén devoted his time to tending his garden and sought solace in the memories which took him back to his wanderings in those landscapes and gardens in China and Japan. The photographs published in *Gardens of China* date to his visits to East Asia in 1918, 1921–3, 1929–30 and 1935.[9] Before examining his employment of the reminiscences in publication, this chapter will explore what is known of his garden through photographs, his articles in Swedish newspapers and a few letters on practical matters such as designs for a garden gate and an arch of Japanese inspiration.[10] Three articles, two of them published in the early years of the war in 1940 and 1941 and the third later in 1958, provide us a view of the principles of the planning of the garden and on the basis of the texts it is also possible to prepare a list of the plants that flourished there.[11]

That Sirén found gardening a solace in turbulent times comes forth in the first paragraph of the 1940 article. It was written in late spring after the first winter of war and the first sentence states bluntly: 'It is difficult to be optimistic this year'.[12] It was hard to prevent thoughts wandering to the drama unfolding in the fields of Somme and Oise. And yet, even if spring was late that year – the first winter of war 1939–40 had been cold and long – he felt that seeing new growth and perseverance had brought a whiff of hopefulness. After all, what is a garden but an education in optimism as we sow and plant our fantasies and memories, hoping that they will sprout and prosper. At the moment of his writing, it was still uncertain which plants had survived the harsh winter and Sirén reflected whether we should draw analogies from this to the current political situation in the world. Nevertheless, his solution was to just focus on the garden and to leave these philosophical reflections to others. Here we can sense an intimation of escapism when he emphasizes that the focus must be on the garden and particularly how it was during happier times.[13] The second article, published close to midsummer 1941, explores poetically the plants – almost animating them – and the atmosphere of the garden, particularly the effects of evening light. Here the presence of war is almost invisible until we come to the end, when he observes that the voices of the stones and music of the flowers are quickly dimmed when the sounds of a world in war reaches them.[14]

The starting point for Sirén, in planning the garden, had been the features and plants which the terrain on his site provided. He denied having followed a particular style of gardening, though admitted that East Asian elements had crept in during the process, step by step. The basic features of his place – barren rocky hill with a few twisted pines in the background – had brought to his mind memories of landscape motifs particularly from Japan. The rocky base provided basins for water features around which more stones were brought in to create a miniature mountain landscape (Figure 5.1).[15]

Later he likened the miniature mountain landscape to Huashan's range and peaks.[16] Viewed from the veranda, the photograph shows the two basins or pools of water and beyond and around them the boulders of different sizes and shapes are placed artfully. Among them, we can see a hut built for a Buddhist statue and also a Japanese lantern. The background is filled with pines, with a few juniper trees here and there.

Basically, rather than plant choices, the presence and arrangement of stones and evergreens created the references to East Asian gardens, particularly to those of Japan. Plants were chosen on the basis that they would thrive in the cold Nordic climate and on this rocky and dry site. The

FIGURE 5.1 Sirén's garden at Lidingö, view from the veranda towards the ponds. Photograph by Sirén 1940, Osvald Sirén Photo Album, Finnish National Library. Image in public domain.

rocks and boulders created a miniature landscape which had the air of being remote and inaccessible. This landscape was complemented by planting. Therefore, there are quite a few alpine plants: *Saxifraga cotyledon*, *Cerastium tomentosum*, *Helicrysum*, *Antennaria*, two species of *Dianthus*, *Pulsatilla vulgaris*, *Cypripedium calceolus*, primulas and a few others. Sirén also noted that after many experiments and mistakes, he had learnt to appreciate plants that embraced and emphasized groundlines of the landscape and covered it with shifting colours.[17]

Nevertheless, a few plants of East Asian origin would thrive in the Nordic landscape. In summer the pond would be home to the water lily accompanied by goldfish, but unfortunately, we do not know whether the water lily was of East Asian origin or of the type from temperate regions of Europe. Sirén noted that these came to their best advantage when viewed from the veranda – playfully called their 'Summer Palace' – in the silvery morning light or in the warm afternoon sun. Nearby the ponds would grow *Iris ensata*. The surroundings recalled the wanderings in Beijing's environs, as if one was wandering along the covered walkways or sitting in a pavilion by a lotus pond.[18]

Elsewhere we could have found a couple of species from the *Trollius* family, *Hypericum henryj*, Chinese lilies, Asian anemones, peonies and azaleas. Sirén's shrubs included several species from the *Cotoneaster* family (*C. horizontalis*, *C. adpressus*, *C. microfylla*, *C. nanshan*, etc.), *Mahonia bealei* and *Berberis candidula*. The cotoneasters were planted near the Japanese lantern. In fact, Sirén maintains that if one has learnt to appreciate both Eastern and Western cultures, one could not avoid bringing the two worlds together in a garden.[19] His main view on gardening is that we should, first and foremost, understand the terrain and the growing conditions (soil, climate, etc.) and then explore what is possible within this particular habitat. This would result in a harmonious landscape. These ideas are reminiscent of thoughts put forward in Ji Cheng's 計成 *Yuan ye* 園冶 (1631), which Sirén had been reading and translating from Chinese.[20] Similar ideas were in vogue in European gardening too, and it is not possible to determine how much came from *Yuan ye* and what from elsewhere. Nevertheless, strict principles of gardening styles had no place in his planning.

Plants were not only appreciated for the colours of their flowers but also for their varied foliage. Sirén did not enumerate all the plants in his garden, though he noted that there was a separate area which he called 'the actual flower garden'.[21] It is interesting to note how Sirén's garden is filled with a great variety of flowers, though he seems reluctant to praise them – in fact he could be quite dismissive of them – and instead highlighted a Japanese stone pagoda which gave the place dignity. The pagoda had been purchased in 1936 through Samurai Shokai, an antiquities and curio shop in Yokohama, owned by Nomura Yozo (1870–1965). Two letters from Nomura in the Sirén Archive deal with the transaction and Nomura is pleased to inform Sirén that he had found 'one very good one, artistic in shape, graceful in form, strong in

FIGURE 5.2 Sirén's garden at Lidingö, the pagoda. Photograph by Sirén 1940, Osvald Sirén Photo Album, Finnish National Library. Image in public domain.

quality and reasonable in price'.[22] A photograph from 1937 shows it newly arrived with space cleared around it but before planting. In summer 1940, the pagoda is well settled in its new surroundings of mostly evergreen planting (Figure 5.2).

A path passes the stone lantern near the pond, leading through a gate (*torii*, as Sirén calls it) assembled from pieces of pine trunks: two verticals topped with a curved one with its ends pointing upwards (Figure 5.3).

This was another project on which Nomura had given Sirén advice, with a comment that 'these rastic [sic] looking gates will always add so much of the feeling to the garden and general landscape'.[23] This gate transformed, according to Sirén, the path into a 'path of virtue' (*dygdens väg*), which is reference to Japanese tea gardens. Sirén notes, however, that in their garden it led instead to a pavilion used for afternoon coffee.[24] There is another gate made of bamboo and of Japanese inspiration on the border of the garden, leading out of it.

The black-and-white photographs convey a sense of sparseness and austerity as we do not see any colours of the flowers. Viewing them while reading Sirén's poetic descriptions shifts the emphasis and reveals the significance of this refuge for him. Figure 5.1 shows the arrangement of rocks behind the two water basins. In fact, Sirén maintained that the positioning of them was inspired by real Chinese mountains such as Huashan

FIGURE 5.3 Sirén's garden at Lidingö, Japanese lantern and the 'torii'. Photograph by Sirén 1940, Osvald Sirén Photo Album, Finnish National Library. Image in public domain.

in Shaanxi Province.[25] Creating a miniature mountain landscape with rocks and plants follows the spirit of tray landscapes and rockeries which a visitor finds in Chinese gardens. Enjoyment of this small-scale scenery is akin to viewing a landscape handscroll and travelling through it in one's mind. This manner of 'travelling while lying down' (*wo you* 臥遊) was emphasized by Chinese painters, most famously Zong Bing 宗炳 (375–443) and Guo Xi 郭熙 (after 1000–*c.* 1090).[26] Sirén was well aware of their ideas regarding landscape, as he had been introducing them in vol. I of his *A History of Early Chinese Painting*.[27]

Sirén carried with him memories of travels in places where he had felt comfortable – in tune with these places. Though his physical home was in Sweden, in some respects he felt that his spiritual home was in East Asia. However, he was not creating a Chinese or a Japanese garden. Neither should we call his garden East Asian. To designate a garden outside China or Japan as Chinese or Japanese is a complex issue and leads easily to garden essentialism and further involves ideas relating to authenticity.[28] In Sirén's garden, the visual references in forms of rock arrangements, objects, gate structures and plants were souvenirs and vignettes he had picked up and which helped him to retain and cherish the connection with the places he had visited.

In his writings Sirén does not make reference to the debate concerning whether non-native or so-called exotic plants should be avoided. This debate evolved around the ideas of natural or wild garden design. However, what is defined as native and what as non-native is often vague and depends also on the geographical location. In the context of this chapter, it might be worthwhile to note that the debate was fierce in Germany and particularly during the time of National Socialism.[29] Exponents of this ideology wanted to purge all exotic plants from Germany, not just foreign peoples. As we have seen in the discussion of which plants he had, in Sirén's setting native and non-native elements and ideas lived happily together.

Finding refuge in memories

While the physical garden increasingly occupied Sirén's time during the war, garden memories of the wanderings in Chinese and Japanese gardens lingered in his mind. These reminiscences were complemented with research and became *Gardens of China*, a book which was published after the war in 1949 (the Swedish edition *Kinas trädgårdar och vad de betytt för 1700-talets Europa: Trädgårdar in Kina* had seen the light of day in 1948).[30] At the core of his refuge in the past wanderings were the hundreds of photographs he had taken during those travels. They were an aid to his visits down memory lane, as were the pictures of Chinese and Japanese paintings which had accumulated over three decades of research. He used images as substitute for physical journeys in times of distress. Here again, Sirén followed the example of Chinese painters who had emphasized the importance of travelling while lying down, as discussed above in connection with the miniature mountainscape.

It is therefore interesting to note that in the introductory chapter to *Gardens of China* ('The Chinese Garden – a Work of Art in Forms of Nature'), Sirén stresses the close relationship between wandering in a Chinese garden and viewing a landscape handscroll. Visitors to a garden are 'led on by ever new impressions, farther and farther into a composition that is never completely revealed, and which for this reason retains something of the secret charm and allurement of the unknown'.[31] Similarly, viewers of a handscroll encounter a succession of unfolding images as they scroll through the painting section by section.

Revisiting his photographs may have been an impulse for Sirén to write about Chinese gardens, but the research resulted in a publication that contains a wealth of factual information and translations from original sources. It covers the basic principles of planning and significant features, considers history of gardens lost and surviving, from Beijing to Jiangnan. His examples range from the imperial and princely to private gardens, with a few examples from temples. One chapter is dedicated to gardens in Japan and some of that material was published already in 1919 in *Den Gyllene*

Paviljongen: Minnen och Studier från Japan (The Golden Pavilion: Memories and Studies from Japan).[32] The study included working through the garden treatise *Yuan ye* and translating it into both Swedish and English.

Occasionally, throughout *Gardens of China*, we get a sense of Sirén revisiting the gardens in his mind as he takes us from one courtyard to another. However, these descriptions of routes through the gardens are quite factual, hardly ever betraying what he felt when looking through the old photographs and choosing ones for the publication. From time to time, though, he does ponder how the gardens might have fared during the current wartime conditions. He reminds the reader that he is writing of the gardens based on what he saw in them a quarter of a century ago and cannot attest to their present condition.[33] Whereas the text does not convey much in the form of personal memories, the photographs themselves are evocative.

Sirén was a skillful photographer: compositions are balanced and carefully planned. In *Gardens of China*, for example, he catches reflections in water (plates 113 and 115) or light rays through doorways (plate 73). Some gardens look desolate and overgrown, such as Gong Wang Fu (plates 134–136) and Cheng Wang Fu (plates 130–131). These images have a nostalgic aura hovering over them. We ought to remember that the 1920s and 1930s, when the photographs were taken, were turbulent times in China and therefore hardly any of the gardens were in pristine condition when he photographed them. Flipping through the plates in *Gardens of China*, we get a sense of the detailed knowledge Sirén had of Chinese and Japanese gardens. Both as a gardener and as a scholar, Sirén escapes in his mind to East Asia, to cultures and locations where he had formed happy and inspiring memories.

This escapism, as has been discussed above, was not confined to wartime; instead, it seems to have become a significant part of the way he tried to accommodate into a life in Sweden in general. The house had been transformed to integrate elements that referred to China and artworks from East Asia decorated the interiors. This process had been initiated in the late 1920s and was accelerated after the return from China in 1930. Similar reshaping was taking place in the garden with the additions of the Japanese pagoda and Japanese-style gates. However, it ought to be emphasized here that Sirén was neither creating a Japanese nor a Chinese garden. He did not discuss his garden in those terms; instead, he admitted that elements referring to East Asian gardens had in a way crept in. I have preferred to stress the souvenir aspect of his collecting practice within the house and tend to see the garden in a similar way.[34] A garden can be viewed as a collection of animate and inanimate objects – plants and man-made things and structures.

Therefore, when the war began, the stage was set for a kind of retreat into a more comforting world of the past. This did not mean that Sirén was inactive by any means. On the contrary, a flurry of activity seems to have taken place. This took the form of publication projects. His magisterial volume of *Kinas Konst under tre Årtusenden* (*Three Thousand Years of*

Chinese Art) had been published in 1942–3 and he was working on a *catalogue raisonnée* of the Chinese collections in the Nationalmuseum.[35]

Would he have been inspired to embark on the research underlying *Gardens of China* and its sequel on the relationship of European eighteenth-century gardens with Chinese ideals, if it had not been for the war and the conditions which kept him tightly in Sweden and in his house and garden? He had to limit himself by researching material that was readily available without embarking on research visits abroad. After all, the history of Chinese art in Swedish and the cataloguing of the Chinese paintings and sculptures in the museum collection were perhaps not the most challenging projects for Sirén at this stage in his life. Whereas a garden history-related topic allowed him to explore further the themes that had appeared in the 1920s publications such as the aforementioned *Den Gyllene Paviljongen* and *The Imperial Palaces of Peking* (1926), which not only discussed the architecture of the palaces but also the gardens within the complexes.

The third newspaper article Sirén wrote about his own garden was published later, in 1958. Its focus was not his garden per se, but the roles of water and stones in a garden. It combines his knowledge of East Asian gardens with his own experiments in bringing water – or sense of water – into his garden, which did not have a water source.[36] We can view this article as a text where his practical experience of gardening comes together with what he had learnt from his research on both East Asian and European eighteenth-century gardening practices. Emphasis is on the role of imagination. Regarding water, Sirén suggests that we should learn from the methods East Asian gardeners used when faced with lack of water. The sense of water could be created by the careful placement of stones and sand. Personally, Sirén had taken advantage of the striations on the granite bedrock created by the ice age as his imagination saw the flow of water in these grooves winding on the rock surface.

Gardening as sanctuary and solace

Winters in the Nordic countries are long and the actual gardening activity on one's plot is limited to few months in a year. That does not mean, though, that thinking of gardening and gardens is similarly limited. A great deal of gardening transpires in imagination when a gardener wades through seed and plant catalogues while planning for the new season, or when they explore publications of more or less famous gardens, watch garden programmes or YouTube videos, or read books on practical advice.

Those who garden know that as spring progresses there is an anticipation about what has survived the cold and what has not. This means preparing oneself for disappointments and this was something Sirén emphasized in his newspaper articles. There is a lot of trial-and-error elements in gardening. Nevertheless, seeing new growth is invigorating. On one hand, gardening is a concrete activity while at the same time it appeals to our feelings. Upon reflection, this is why gardening can be so soothing in times of distress, whether it is wartime, time of bereavement, or an aftermath of traumatic experiences. It allows for escape from the real world, so to speak, and yet it anchors us to a hands-on occupation.

There is useful research and literature on gardening in distressing times and situations, on the way gardening helps us to calm our minds and to get through traumatic experiences. A rewarding and eminently readable study is Sue Stuart-Smith's *The Well Gardened Mind* and similar aspects are also covered in George McKay's *Radical Gardening: Politics, Idealism & Rebellion in the Garden*.[37] Kenneth Helphand's *Defiant Gardens: Making Gardens in Wartime* focuses specifically on gardens of war in the twentieth century.[38] Helphand covers gardens which have since disappeared: those made by soldiers in the trenches, by inhabitants in ghettos and in internment camps. These were gardens made in extreme living conditions and therefore are a poignant contrast to Sirén's gardening. The dangers and effects of war were not as imminent and concrete for Sirén as they were for those who gardened in the trenches or war zones.

It has been somewhat disconcerting to research Sirén and his life during the Second World War – not because of the horrors of war, but because there is so little visible of those horrors in the material that survives. Was he trying to avoid thinking that the world was in war? Maybe it was easier when one was not constantly threatened by bombings. The references to the world situation are brief in the aforementioned articles and he is quick to move on to write about the garden and its delights. The photographs taken there during 1940–3 do not differ from those of the 1950s. In all of them the compositions are balanced, the light glimmers and it is summertime. There are no human beings and no visible decay or neglect. Nothing points away from the idyllic views, that there might be more troubling events happening beyond the garden's borders.

Here we might make a connection to pastoral as a nostalgic retreat into the past. Helphand brings up this aspect and sees there 'the danger that the garden becomes a trite palliative, a mask, and that working in the garden reflects an abdication of involvement and responsibility. At the extreme, the garden seems cowardly'.[39] Nevertheless, he is aware of and emphasizes the paradox ingrained in the pastoral. It is not necessarily about escape from the current situation and can, instead, be viewed as something positive: 'the garden is not only a calm retreat and welcome respite, but also the assertion of a proposed condition, a model way of being and of physical form'.[40]

Nevertheless, the point on how gardening can be a place of refuge for the mind and an experience which gives hope for a better future is characteristic for all these situations, whether one was in the midst of war action, at the home front, or in a country endeavouring to stay neutral. From more recent times, we could make a connection to the COVID-19 pandemic of the early 2020s in the sense that one's world shrinks as the extent of movement becomes restricted and daily life becomes repetitive. In a situation like this, it is easy to turn inward, to find solace in memories, or to explore the world through literature or artworks (if one has access to them), and of course in the current world there is the internet and what it offers. A garden is a safe haven for both dreaming and getting to grips with the real world. It is perhaps, in reference to gardening, appropriate to quote a passage from Sue Stuart-Smith in this context:

> Like a suspension of time, the protected space of a garden allows our inner world and the outer world to coexist free from the pressures of everyday life. Gardens in this sense, offer us an *in-between* space which can be a meeting place between our innermost, dream-infused selves and the real physical world.[41]

It is interesting to compare the photographs of Sirén's garden with the ones of European gardens reproduced in *China and Gardens of Europe of the Eighteenth Century*, the companion volume to *Gardens of China*.[42] After the war was over and travel was again possible, Sirén toured France and England photographing eighteenth-century gardens and estates. Though the images he captured are as poetic and beautiful as always when he has been behind the camera, it was not entirely possible to avoid the inclusion of decay and devastation in the scenes. After all, many of the places had faced neglect for years: plants had gone wild and buildings had started to fall apart. Therefore, it was disconcerting to read one book review where the writer complained about the inclusion of so many photographs depicting ruinous buildings and landscapes.[43] As if that could have been avoided.

Désert de Retz (original name Désert de Monville) was almost completely ruined when Sirén visited it after the war, though there the neglect had persisted longer, since its days of glory before the French Revolution.[44] Many of the buildings in this landscape were actually designed as ruins, though not the Chinese House (Maison Chinoise which no longer exists). It was in its heyday a beautifully decorated piece of French Chinoiserie architecture, but in Sirén's photographs it appears as if it might collapse any minute.[45] Similarly ruined buildings were found also in England, for example, in Painshill Park, where the Temple of Bacchus has lost some of its columns and the tympanon is supported by a wooden beam and the walls are full of holes.[46]

Returning to Sirén's own garden, the photographs raise the question of where were the vegetables? The newspaper articles mention apple trees, but nothing about the wartime necessity of turning flowerbeds into vegetable plots, as was happening elsewhere. In Britain this was phrased as 'digging for victory'.[47] There, for example, even famous gardens such as Sissinghurst suffered and became overgrown when there were no gardeners to tend them as before.[48]

At the beginning of this chapter, I quoted the first lines of Vita Sackville-West's poem *The Garden*. The poem contains several references to gardeners and their life during those distressing times. She also couches the everyday work of a gardener in war-like vocabulary, when she writes:

> So in the gardener's more persistent war
> Where man not always is the conqueror,
> We plodded as we could, and fought
> Permanent enemies, of weed and wing:
> The strangling bindweed and the running strands . . .[49]

But she also notes in the poem on how flowers felt lovelier for the danger.[50]

Notes

1 Osvald Sirén to Florence Ayscough, 29 June 1930, file A:3/280, Sirén Archive, Museum of Far Eastern Antiquities, Stockholm.

2 The Finnish-Swedish art historian Osvald Sirén was the focus of my research in 2005–15, resulting in a monograph *Enchanted by Lohans: Osvald Sirén's Journey into Chinese Art* (Hong Kong: Hong Kong University Press, 2013) and several articles. In this chapter, I am revisiting the material in order to take a renewed look at his life during the Second World War and to reflect on how people cope in politically turbulent times or negotiate times of adversity.

3 Vita Sackville-West, *The Garden* (London: Frances Lincoln, 1946), 13.

4 For detailed information of Sirén's career and publications, see Törmä, *Enchanted by Lohans*.

5 Osvald Sirén, *A History of Early Chinese Painting*, vol. I (London: The Medici Society, 1933), 69–70.

6 Ibid., 121.

7 Osvald Sirén, *A History of Early Chinese Painting*, vol. II (London: The Medici Society, 1933), 82.

8 I have discussed Sirén's interiors and object displays extensively in ch. 4 in my *Nordic Private Collections of Chinese Objects* (London and New York: Routledge, 2020).

9 Sirén did not visit China between 1935 and 1950, so none of the photographs were taken during the Second World War.

10 The photographs cover a period from 1933 to 1951 and are from an album in the collection of the National Library of Finland in Helsinki. There are also photographs of Sirén's home and garden in the Sirén Archive at the Museum of Far Eastern Antiquities in Stockholm.

11 Osvald Sirén, 'Om en trädgård', *Svenska Dagbladet*, 2 June 1940; Osvald Sirén, 'Juninatt i trädgården', *Svenska Dagbladet*, 28 June 1941; Osvald Sirén,

'Lägga sten och leka vatten', *Svenska Dagbladet*, 24 August 1958. The 1958 article is discussed in the conclusion.

12 Sirén, 'Om en trädgård'. The original Swedish reads: 'Det är svårt att vara optimist i år'. All translations are mine unless otherwise noted.

13 Sirén, 'Om en trädgård'.

14 Sirén, 'Juninatt i trädgården'.

15 Sirén, 'Om en trädgård'. The information on plants is repeated in Sirén, 'Juninatt i trädgården'.

16 Sirén, 'Juninatt i trädgården'.

17 Ibid.

18 Ibid.

19 Ibid. Sirén refers in Swedish to 'Öster- som Västerland'.

20 Sirén's translations were published in Osvald Sirén, *Gardens of China* (New York: Ronald Press Co., 1949). Sirén used as his source for the Chinese text reprint: Beiping, Zhongguo yingzao xueshe, 1932.

21 Ibid: 'den egentliga blomsterträdgården'.

22 Nomura Yozo to Osvald Sirén, 18 June 1936, Sirén Archive, Museum of Far Eastern Antiquities, Stockholm.

23 Nomura Yozo to Osvald Sirén, 17 September 1937, Sirén Archive, Museum of Far Eastern Antiquities, Stockholm.

24 Sirén, 'Om en trädgård'.

25 Sirén, 'Juninatt i trädgården'.

26 For English translations of their texts, see Susan Bush and Hsio-yen Shih, *Early Chinese Texts on Painting* (Cambridge MA and London: Harvard University Press, 1985), 36–8 (Zong Bing) and 150–4, 156–60, 165–9, 177–82 (Guo Xi).

27 See footnote 5.

28 On the issue of garden essentialism and how we might discuss this type of cultural translation, see Christian Tagsold, *Spaces in Translation: Japanese Gardens and the West* (Philadelphia: University of Pennsylvania Press, 2017), 85–94.

29 Joachim Wolshke-Bulmahn, 'The Nationalization of Nature and the Naturalization of the German Nation: Teutonic Trends in Early Twentieth-century Landscape Design', in *Nature and Ideology: Natural Garden Design in the Twentieth Century*, ed. Joachim Wolshke-Bulmahn (Washington DC: Dumbarton Oaks, 1997), 187–99.

30 Sirén, *Gardens of China.*

31 Ibid., 4.

32 Osvald Sirén, *Den Gyllen Paviljongen: Minnen och Studier från Japan* (Stockholm: P. A. Norstedt & Söners Förlag, 1919).

33 See for example Sirén, *Gardens of China*, 93.

34 See ch. 4 in Törmä, *Nordic Private Collections.*

35 Osvald Sirén to Jean Buhot, 20 October 1945, Sirén Archive, Museum of Far Eastern Antiquities, Stockholm.

36 Sirén, 'Lägga sten och leka vatten'.

37 Sue Stuart-Smith, *The Well Gardened Mind* (London: Williams Collins, 2020); George McKay, *Radical Gardening: Politics, Idealism & Rebellion in the Garden* (London: Frances Lincoln, 2011). To these we could add Charles A. Lewis, *Green Nature/Human Nature: The Meaning of Plants in Our Lives* (Urbana, Chicago and Springfield: University of Illinois Press, 1996).

38 Kenneth I. Helphand, *Defiant Gardens: Making Gardens in Wartime* (San Antonio: Trinity University Press, 2006).

39 Helphand, *Defiant Gardens*, 226.

40 Ibid., 228.

41 Stuart-Smith, *The Well Gardened Mind*, 16.

42 Osvald Sirén, *China and Gardens of Europe of the Eighteenth Century* (New York: Ronald Press Co., 1950).

43 Helen M. Fox, 'Chinese Influence of European Gardens', *Journal of the New York Botanical Garden*, October 1950.

44 Sirén, *Gardens of Europe*, 115.

45 Ibid., plates 88–89.

46 Ibid., plate 35.

47 Jane Brown, *The Pursuit of Paradise: A Social History of Gardens and Gardening* (London: HarperCollins, 1999), 263–4.

48 Matthew Dennison, *Behind the Mask: The Life of Vita Sackville-West* (London: William Collins, 2014), 263–4, 269.

49 Sackville-West, *The Garden*, 97.

50 Ibid., 90.

PART III

Moral virtues and poetic virtues

6

Garden practices, 'quiet' virtues and understanding

David E. Cooper

Gardens, moral purposes and virtues

From ancient times to the present, a wide variety of broadly moral purposes have been attributed to the garden and to the practice of gardening. A Chinese Song dynasty official, for instance, claimed that enjoyment of a pavilion and its surroundings would 'moralise people' more effectively than preaching to them could achieve.[1] To take a more recent example, each year at the Chelsea Flower Show these days, visitors are guaranteed to encounter 'eco-gardens' whose aim is to encourage people to 'save the planet' and forestall the suffering that will be caused by global warming and related developments. A few years ago, I was shown, by its creator, a 'guerrilla' garden in Münster whose purpose was to arouse in well-off Germans a sense of guilt for the extreme poverty found in sub-Saharan Africa.

As these examples indicate, there has indeed been variety among the many claims that have been made about the garden's moral purpose. The purposes themselves have varied from the encouragement of austere self-discipline to the reinforcement of Victorian values of industry and family life. They have ranged from inspiring a patriotic love of one's country to promoting senses of justice and community. The popular Alnwick Garden, close to where I live in the north-east of England, lists among its objectives the well-being, education and empowerment of the 'most vulnerable community members', and inspiring people to become 'changemakers' equipped to address 'the climate crisis'.

Equally varied have been the means that have been proposed for how gardens may contribute to moral purposes. These have ranged from a didactic conveyance of moral lessons to a rather more subtle exemplification of moral ideals. An example of the former would be the temples in Stowe

Park (Plate 2), one of them containing statues to commemorate ancient virtues, the other a fake ruin designed to convey that, by the eighteenth century, there were no virtues left to celebrate. An example of the latter would be the way in which the informal, 'natural' landscapes of 'Capability' Brown, also created in the eighteenth century, were held to exemplify the love of liberty allegedly prized by Englishmen. Formal French gardens, by contrast, were deemed to exemplify a desire to control and dominate nature – to be 'tokens of her slavery', as Schopenhauer put it.[2]

Other gardens were held to promote morality by providing apt environments for moral reflection, while yet others were intended to instil a 'work ethic'. It was the belief of some Victorian writers and gardeners that vigorous gardening might divert young men and women from their preoccupation with sex.[3] At the Alnwick Garden, mentioned earlier, moral objectives are variously pursued through education courses, a 'community fridge' to encourage responsible food practices, 'climate action courses', and a 'poison garden' to remind visitors of the caution with which they should treat natural environments.

It should be clear from the examples I have given that by 'moral purposes', I do not simply have in mind those that are understood – as they tend to be nowadays – in terms of moral principles, rights and duties. It is with a broader – and older – conception of the moral or the ethical that I am concerned: one that refers to the good life – to a fulfilled and flourishing human life. In this conception – shared by the moral thinkers of ancient Greece, China and India – 'the point of entry for moral reflection' is a person's concern for the kind of life that they should be leading.[4] In this chapter, I shall focus upon an especially important – and certainly a philosophically interesting – form that claims about the moral purpose of the garden, in this broader sense, have taken in several traditions. This is the idea that gardening and other garden practices may help to cultivate virtues, that they have a significant role in, one might say, the education of the virtues. By 'virtues', I mean those dispositions of character that contribute positively – or at any rate are believed to do so – to leading a good and flourishing life.

Before I expand on and examine this idea, a couple of points are worth making. First, the moral purposes I refer to should be distinguished from the many therapeutic goals that have been assigned to gardens and gardening – improvement of one's health, for example, or stress reduction, or a remedy for loneliness.[5] That said, the distinction here is not sharp. For Aristotle and Confucius, for example, friendship is an ingredient in the moral life: one ought to have friends, in which case a cure for loneliness achieves a moral, as well as a therapeutic, aim. Second, I sometimes refer to garden practices, rather than simply to gardening, since while gardening may be the most central and obvious of the practices conducted in gardens, it is not the only one. Other practices engaged in by gardeners include designing parterres or pergolas, encouraging wildlife into the garden, placing ornaments, seats and so on in their gardens, and entertaining friends *al fresco*.

The idea I want to discuss – that garden practices may cultivate virtues – is, to be sure, a very old one, as old as the more general thought that gardens can have a moral purpose. The Roman writer Pliny the Younger's ambition, in creating his gardens, was to 'cultivate [him]self' and lead 'a good and genuine life'.[6] The idea is very prominent, too, in the great East Asian traditions of the garden as a 'Way' (Ch. *dao*, Jap. *dō*). A Way, as it is understood in these traditions, is not a mere recreation, hobby or activity, but a practice of self-discipline, self-development and, in one authority's words, 'bio-spiritual cultivation'.[7] The idea I am discussing is, however, not confined to the Ancients: it is still very much with us in our own times. Consider, for instance, the popular belief that good gardening may foster an appropriate attunement and comportment to the natural world. Or consider, too, the attempts sometimes made to use gardens as places where troubled young people – ones with a drug or alcohol problem, say – might recover a sense of the value of work and communion with other people. (Monty Don, who was responsible for one such project, provides a moving account of it in his book *Growing out of Trouble* [2006].)

'Internal' and 'quiet' virtues

The alleged virtues that garden practices have been held to cultivate are, as already indicated, very varied: patriotism, self-discipline, environmental responsibility, diligence, industry, and many others. The question that I want to address is whether there are some virtues that are, as it were, 'internal' or 'intrinsic' to garden practices, while others are at best only contingent or 'accidental' ingredients or products of these practices. Are there, to put the question differently, some virtues that might be said to belong to the 'essence' of gardening?

It would be possible to design a garden as a kind of assault course for training soldiers and, hence, for instilling physical courage. Wealthy Englishmen once designed gardens that contained outdoor theatres, with the aim that these would encourage aesthetic appreciation of music or drama.[8] French and Italian gardens were sometimes intended to provide spaces ideal for romantic trysts. But we can agree, I think, that these are functions that are imposed on the garden, and that gardening *per se* is not especially or characteristically apt for encouraging courage, musical appreciation or sexual love. The same could be said about the use of gardens, as with the temples at Stowe, to convey moral messages through statues, inscriptions on stones, or whatever. I would say the same, too, of, the intentionally scraggy, almost plantless and sandy 'guerrilla' garden in Münster that I referred to earlier. The garden, which contrasted vividly with the richly planted park in which it stood, was designed by its maker to prick our conscience about world poverty. Perhaps it did so, and quite effectively, but it is not intrinsic to garden practices to encourage a compassionate concern for disparities in wealth.

There are, in short, all sorts of ways in which gardens might be employed to encourage virtues or desirable attitudes – including several that I have already mentioned – but not all of these reflect, one might say, the essence of the garden or gardening. Not all of them, that is, are virtues that gardening by its very nature is liable to foster. By contrast, consider such virtues as patience, humility, care, hope, and what the Chinese call *jing*, a sort of 'respectful reverence' for things. These, I propose, are internal to garden practices in a way that courage, musical appreciation and sympathy for the poor and hungry are not. This means, roughly speaking, that they are qualities that are necessary to good gardening, essential to the success of the gardener's endeavour. More exactly, they are at any rate qualities without which the gardener would have to be very fortunate to make and maintain a garden that flourishes – a garden, for example, in which the flowers beautifully blossom, the vegetables are tasty and nutritious, and the birds feel safe and welcome. One can at least imagine that, in the location of a particular garden, the weather, the soil, the wildlife and surrounding environment are so unusually benign and favourable that even impatient, careless, irreverent and overly self-confident gardeners manage to grow tasty carrots or beautiful irises. But, then, they are indeed lucky, for the success of the garden owes little or nothing to them and nearly everything to good fortune and benevolent circumstances.

Under normal circumstances, however, a virtue like humility – by which I mean a proper appreciation of one's dependence on conditions not under one's control – is essential to successful gardening. A gardener who ignores changes in the seasons, weather patterns or soil quality does so at their considerable peril. The same is true of a virtue like care – a concern for the good of the plants, trees, or creatures with which the gardener engages. Bad gardeners may neglect to prune and protect their fruit trees, but they will then be very fortunate if, despite this, the trees produce plenty of fruit.

Gardening, it goes without saying, is a practice, and therefore requires skill and discipline. Its internal or intrinsic virtues are those that are normally required for the proper exercise of this skill or discipline, of this 'Way'. This way of understanding the notion of virtues as internal to a practice was first articulated by Alasdair MacIntyre in his seminal book, *After Virtue* (1982). In the case of, say, judo and other martial arts, these virtues will include courage, determination and respect for one's opponent. These are the virtues internal to such arts: the timid, hesitant yet disdainful fighter will be very fortunate to win a contest. In the case of garden practices, it is, instead, the 'quiet' virtues – of care, humility and so on – that are paramount. Or, so I am proposing.

Why, however, do I refer to them as 'quiet' virtues? The label strikes me as an appropriate one for two reasons. For a start, they are virtues that contrast with the more muscular, more strident virtues usually associated with, and perhaps necessary to, what is nowadays regularly called social and political *activism*. The aim of achieving global equality, say, or the objective of 'saving the planet', or the attempt to get rid of a corrupt political regime are goals that may be furthered by, and in turn encourage, such putative virtues as righteous anger, a burning sense of injustice, and concern for the well-being of one's grandchildren. Activism, in this sense, stands in contrast with what is often referred to these days as *quietism*. The quietist has gentler, more personal and less politically driven priorities. The virtues internal to garden practices, I am suggesting, tend towards the quietist, not the activist, end of the spectrum of moral ambitions. ('Quietism' in this sense, one should note, is distinct from some other uses of the term – as the name, for example, of a seventeenth-century theological tendency in Spain and France, or as a label for the style of anti-metaphysical philosophical investigation associated with Ludwig Wittgenstein.)

A more important reason for referring to the quiet virtues is the intimate association they have with a number of ancient and traditional ideals that may themselves be characterized as forms of quietude. Among such goals or ideals are the *ataraxia* and *apatheia* – the freedoms from disturbance and the passions – admired by the Stoics, Epicureans and other Hellenic Greeks. Further east, we encounter the goal of equanimity (*upekkha*) that is one of the 'sublime states' (*brahmavihara*) which the Buddha exhorts his followers to seek. Further east still, we encounter the *wu wei* ('not doing', 'inaction', or 'effortless action') prized by the Daoist sages. What all of these conceptions share is the praise of a way of living that is tranquil, at ease, undisturbed – a life that is 'happy, calm, and unshaken', as one garden lover, the Roman Stoic, Seneca, put it.[9] The quiet virtues, then, are those that are cultivated not least in order to achieve and secure quietude and related conditions, such as tranquillity and equanimity, in one's life. Better, perhaps, they are cultivated as essential *constituents* of such a life.[10] Tranquillity, for instance, is secured in and through the exercise of humility and adjacent virtues. 'Quiet virtues' seems like an apt title for virtues that conduce to conditions marked by quietude and its cousins.

'Refuge' and understanding

Due recognition of the quiet virtues intrinsic to garden practices is helpful, in my judgement, in explaining a number of enduring perceptions, stretching back once more to ancient times, of the significance of gardens. This is a reason to attend to these virtues. I shall focus on two of these ancient perceptions: that of the garden as a place of retreat or refuge, and that of gardening as a source of understanding, wisdom even.

If the quiet virtues are internal to garden practices, then one of the several senses in which the garden has often been considered as a place of retreat or refuge becomes readily intelligible. Gardens have served, very occasionally, as refuges in a quite literal sense. They have been places where people have sought physical protection or total disassociation from the outside world. More commonly discussed, however, has been the role of the garden as a place in which to find temporary relief from 'the dust and grime of the world',[11] as several Chinese texts call it, or to enjoy what Andrew Marvell in his famous poem, 'Thoughts in a Garden', called 'fair quiet' and 'delicious solitude'.[12]

More interesting, however – at least in the context of my discussion – is a more figurative or metaphorical image of the garden as refuge. I have in mind the idea of the garden as a space in which it is possible to cultivate and exercise virtues that are remarkably difficult, or even impossible, to practise and develop in the everyday, outside world of work, business and social intercourse. Andrew Marvell lists 'innocence', 'repose' and 'fair quiet' – things for which there is little opportunity in the busy world of everyday life – among the rewards that a garden affords. Now, the quiet virtues are surely ones whose exercise in that world is also a considerable challenge. For this is an aggressive, often febrile and agonistic world that emphatically discourages, even punishes, humility, say, or forbearance and patience. Assertiveness, commitment, passion, self-confidence and ambition seem to be the qualities necessary to fit into that world. There is no time or place in it for the quiet virtues in the marketplace, think-tank or busy office.

It is the kind of world from which the Chinese literati, as much as Marvell, sought refuge in order to conduct, as they saw it, better lives, ones in which such virtues could more feasibly be exercised.[13] For Pliny, too, it was clearly the hospitality of the garden to the quiet virtues – ones that he knew were hard, even dangerous, to exercise on the streets of Rome – that made it a place conducive to a 'good and genuine' life. Refuge is sought, for these writers, from what Ian James Kidd calls 'the moral harms' to life in the societies in which they wrote.[14] On this perception, the garden, one might say, is seen as a refuge *for* the virtues – for, in particular, those quiet virtues to which the everyday world is so inhospitable. In the garden, these virtues have the space and time in which to breathe.

An emphasis on the quiet virtues also helps to explain the ancient perception of the garden as an arena for the cultivation not only of moral virtues, but of an understanding of the scheme of things – of reality and one's place in it. The garden, as Robert Pogue Harrison puts it,[15] has long been a means for 'disclosing' the world. This was an understanding that, in contrast to merely theoretical or technical knowledge, was held in ancient traditions to constitute wisdom. Epicurus's garden in Athens, as Harrison points out, was intended to provide 'a form of [philosophical] education'.[16] Medieval gardens in Christian Europe were designed to intimate or illustrate religious truths. The distinguished contemporary Spanish gardener Fernando Caruncho speaks for many other gardeners, I suspect, in celebrating the 'spiritual and cognitive experience'[17] afforded by at least some gardens (Figure 6.1).

FIGURE 6.1 Pazo Pegullal in Salceda de Caselas, Vigo, Spain designed by Fernando Caruncho. Photo by Raúl de Tomás. Courtesy of Caruncho Garden & Architecture.

There should be no tension or competition between emphases on both the moral and the epistemic benefits of gardening respectively, on its contributions both to the good life and to wisdom. The contrasts, here, are anyway fuzzy ones. There is no clear point at which a concern for living well diverges from one for knowledge and understanding. Thinking about – or 'interrogating', as the British gardener and author Alasdair Forbes puts it[18] – the way we should live is one important way, after all, in which living well may be fostered. And it is only in contexts that are morally favourable that the pursuit of wisdom may be possible. In an oppressive society where, say, challenges to religious orthodoxy are severely punished, the accumulation and transmission of wisdom would itself be a considerable challenge. In the 'Ways', referred to earlier, that have played such a significant role in East Asian cultures, moral development and understanding of the world and one's place in it are, as I've argued elsewhere,[19] inextricably combined.

It is worth elaborating here on the mode of understanding or knowledge that, for Epicurus, Caruncho and others, the garden is apt for promoting or cultivating. What is not intended is the sort of propositional, empirical – perhaps technical and scientific – knowledge that might best be acquired through aggressive debate and excitable exchanges of ideas in a think-tank seminar or a brainstorming session. It is not the type of knowledge that could be arrived at by piling up data and evidence, or by undertaking a number of 'research projects'. The understanding aimed at and fostered in Epicurus's garden, for example, was of 'the harmonies of the cosmos' and the soul's relation to matter.[20] The Daoist garden, wrote Maggie Keswick, aimed in its quiet way to render more salient 'the inexpressible experience of the *dao*'.[21] And, as Yuriko Saito nicely puts it, a Zen garden seeks to provide a 'glimpse of the world as it appears to a Zen-enlightened sensibility' (Figure 6.2).[22]

FIGURE 6.2 Dry Garden in Ryoanji. Photo by Stephane D'Alu, CC BY-SA 3.0, via Wikimedia Commons.

These kinds of understanding do not consist in acknowledgement of the truth of certain propositions or beliefs. This is not to deny, of course, that the good gardener will need to know lots of *facts* – about propagation, soil acidity, weather patterns, insect control and so on. Nor, obviously, is it being denied that the good gardener requires a considerable stock of 'know-how': they must, of course, have the skills that are needed when pruning, grafting, planting and designing. But the kinds of understanding I want to identify are not mere bits of information that a person might come across, nor are they individual 'knacks' and practical skills. Rather, they take the form of experiencing or sensing reality, or the way of things, for what it is. This is a form of philosophical understanding unlikely to be available to the arrogant, hubristic, impatient person absorbed in their daily business and engagements. Nor, indeed, is it available to people absorbed in the pursuit of technical or 'calculative' knowledge. It is a kind of understanding, rather, that is available only to the patient, receptive person who is capable of suspending or 'bracketing' everyday practical concerns, and who is willing to respond to things as they are. As one scholar of Chinese philosophy expresses it, when describing the Daoist ideal of *zi ran* ('spontaneity' or 'naturalness'), the understanding sought is 'responsiveness in the impersonal calm when vision is most lucid'.[23]

It is not difficult to appreciate that the kind of understanding of which we have been speaking – the kind that the garden invites – is precisely the type accessible to people who possess the quiet virtues that are cultivated in the garden. It is because these virtues are exercised in the garden that gardening is a vehicle for this kind of understanding, this 'form of education'. Philosophical understanding, or wisdom, requires, as the ancients well knew, the exercise of humility, patience, respect, mindfulness and other quiet virtues. This is why, for example, in the Buddha's eightfold path to enlightenment, moral and epistemic virtues are integrated. Wisdom, unlike the forms of knowledge that are so prized in modern societies, presupposes an openness, stillness and clarity – a sort of 'innocence' – that, as Martin Heidegger (1977) eloquently laments, has little place in the everyday world of 'calculative', technological thought.[24] The garden promises to provide a place in which these qualities might be nurtured.

Gardens and nature

A question that readers who may be sympathetic to my conception of garden practices as cultivating the quiet virtues and philosophical understanding might nevertheless want to raise is the following: Are garden practices at all unique or even distinctive in doing this? Why, if at all, these readers are asking, is gardening especially apt for promoting these virtues and this understanding? Couldn't other practices – and other forms of retreat and refuge from that of the garden – be equally apt and effective? Might not retreating, like Marcel Proust, into a cork-lined room in Paris and writing an enormous novel be similarly apt? Might not driving across the USA, like Ken Kesey, in a converted, technicolour bus with some fellow hippies be equally effective?

I do not want to be dogmatic here and implausibly insist that only the garden can serve as a theatre for the cultivation of the quiet virtues and understanding of the way of things. But I do want to suggest that it is an especially apt space or refuge for this purpose. And I do so despite a couple of worries that deserve an airing. First, we should recognize the force of Kidd's point that in the modern, far from innocent world, it may be difficult even in the garden to find refuge from the 'moral harms' that multiply beyond its walls. Like any would-be refuge, he concludes, the garden may be unable to 'protect itself against [the] world' beyond.[25] But, as that way of putting it suggests, the garden is at any rate no worse off than any other putative refuge, and the possibility that is a distinctively apt place of refuge remains.

Second, we should consider the charge that my remarks about garden practices are touchingly out of date. Gardening, I argued, cultivated virtues and understanding through their contribution to the flourishing of gardens. For example, humility, patience and spontaneity – in the senses indicated – were required, I proposed, if a garden is to prosper. But, someone will ask, are they still required these days when countless aids – fertilizers, frost-proof bulbs, electronic pest deterrents, under-soil heating, sprinkler systems and so on – can be purchased from a large garden centre and used to minimize the chances of even a clumsy and impatient person's garden failing? The worry that technological 'contrivances' might corrupt the work of the gardener is an ancient one. The Daoist sage Zhuangzi describes a peasant who angrily rejects the offer of a 'tricky contrivance' for irrigation, since such devices make people lose contact with 'the pure and simple . . . and vital'.[26] This kind of worry, of course, is not confined to gardening. In many spheres – in sports and the arts, for instance – the need for traditional virtues and forms of understanding is affected by the availability of technology and its 'contrivances'. There is no reason, therefore, to conclude that gardens do not remain especially apt places for the cultivation of virtues and understanding, even if the garden practices of some people are being modified by technology. Anyway, it is important not to exaggerate the worry. Most of us, particularly if we are not especially rich, still have to exercise care, patience, humility and the rest of the 'quiet' virtues if our vegetables and flowers are to grow well.

If, despite these worries, we remain with the claim that the garden is an especially apt place of refuge, it is reasonable to ask why this is so. The reason must surely – and perhaps unsurprisingly – lie in gardening's engagement, and manner of engagement, with *nature*. It must be due to its intimate dealings with natural environments, creatures, plants and other natural entities. It is no accident that Epicurus refers to the understanding of, and comportment towards, nature that he teaches as being, in the first instance, cultivated in the garden. Nor is it surprising that Daoist-influenced poets, such as Tao Yuanming (Plate 3) around the turn of the fifth century, in their search for a 'spontaneity' (*zi ran*) that accords with the way of nature – indeed, with the *dao* itself – should have sought this in gardens.

The relationship between culture and nature is, of course, a large theme and I shall only touch upon a couple of its aspects. The English philosopher Mary Midgley criticized Existentialist thinkers, especially Jean-Paul Sartre, for regarding the world as so much 'dead matter' on which we allegedly totally free creatures may impose our wills and 'projects'.[27] Had they taken notice of the 'biosphere' – the world of plants and animals – they could not, she argues, have adopted this hubristic vision of the unconstrained freedom of human beings. She implies, furthermore, that their failure to pay proper attention to the living world was due to their spending their lives in bars, cafés, jazz clubs, cinemas and other quintessentially urban and all-too-human venues. Midgley's message seems to be that if Sartre – who indeed hated being in the countryside, away from the city – had done a good day's gardening, he would soon have appreciated our dependence on a recalcitrant natural world and the severe limits against which our 'projects' run up. More generally, Midgley is implying, it is essential to the cultivation of the quiet virtues – and even to recognizing that they *are* virtues – to spend time in natural environments, and to engage with them in the manner, for example, of the gardener.

But, secondly, if garden practices, through their intentional and inevitable engagement with nature, help to teach us lessons of humility and forbearance, of patience and hope, they also serve to inspire care and regard for nature. The good gardener learns not only that they depend for their success on the cooperation of nature, but that success is impossible without attending to the good of the plants, soil and creatures with which they engage. The gardener learns, in effect, to honour the duties of care and solicitude that ought to accompany our intrusion into the natural world. When we bring nature into the orbit of culture, art and utility, we should, at the same time, bring it into the sphere of moral concern. If the garden is a forceful reminder of our dependence on what is beyond our control, it is also a symbol of nature's dependence on – its susceptibility to – human ambitions and projects. Gardening is, perhaps, the most striking exemplification one can find of the mutual dependence – the reciprocal inflection and 'porousness', to use Alasdair Forbes's term – of human cultural practices and nature.[28] And this is why, I propose, it is in a privileged position to invite, employ and foster the quiet virtues.

Notes

1 K. Li, 'Landscape Improvement and Scenic Sites in Pre-modern China: A Critical Review' (PhD, University of Sheffield, 2005), 138.

2 Arthur Schopenhauer, *The World as Will and Representation* (New York: Dover, 1969), 405.

3 Keith Thomas, *Man and the Natural World: Changing Attitudes in England 1500–1800* (London: Penguin, 1984), 234.

4 Julia Annas, *The Morality of Happiness* (New York: Oxford University Press, 1993), Ch. 1.

5 See Lucy Jones, *Losing Eden: Why Our Minds Need the Wild* (London: Penguin, 2020), Ch. 5 and Isis Brook, 'Gardening and the Power of Engagement with Nature for Mental Well-being', in *The Oxford Handbook of Mental Health and Contemporary Western Aesthetics*, ed. Martin Poltrum et al. (Oxford University Press, 2023), for two recent surveys of 'horticultural therapy'.

6 Pliny, *The Letters of the Younger Pliny* (Harmondsworth: Penguin Books, 1963), 43, 112.

7 Russell Kirkland, *Taoism: The Enduring Tradition* (New York: Routledge, 2004), 33.

8 The Chinese literati had a long tradition of designing gardens that contained outdoor theatres or performing theatres in a garden; see Lin Zhu's chapter, this volume.

9 Seneca, *Moral Letters to Lucilius* (London: Heinemann, 1917–25), Letter 59.

10 For a case study on the cultivation of these virtues in the gardens created in seventeenth-century Holland and England and their reception of ideas of Confucianism and Chinese gardens, see Yue Zhuang, 'Temple, Huygens and "*sharawadgi*": Tempering the passions to achieve tranquillity', *Studies in the History of Gardens & Designed Landscapes*, 41, no. 4 (2021): 288–308.

11 Tao Yuanming, 'Returning to Live in the Country', in *Tao Yuanming: The Complete Works*, trans. Earl Trotter (Chatham ON: Peach Blossom Press, 2022), 54.

12 Andrew Marvell, 'Thoughts in a Garden', in *The Oxford Book of English Verse, 1250–1918*, ed. Arthur Quiller-Couch (Oxford: Oxford University Press), 402. On these and other testimonies to the garden as refuge, see Ian James Kidd, 'Gardens of Refuge', *Daily Philosophy* (2021), https://daily-philosophy.com/kidd-gardens-of-refuge, accessed 6 October 2023, and his 'Gardens of Refuge, Innocence, and Toil', this volume.

13 See Hongde Wu and Stanislaus Fung's chapter on a well-known Chinese literati garden, Zhuo Zheng Yuan (the Garden of Artless Administration), this volume.

14 Kidd, 'Gardens of Refuge', this volume.

15 Robert Pogue Harrison, *Gardens: An Essay on the Human Condition* (Chicago: The University of Chicago Press, 2008), 124.

16 Ibid., 73–4.

17 Fernando Caruncho, 'The Spirit of the Geometrician', in *Vista: The Culture and Politics of Gardens*, ed. Noel Kingsbury and Tim Richardson (London: Frances Lincoln, 2005), 114.

18 Alasdair Forbes, *On Psyche's Lawn: The Gardens at Plaz Metaxu* (London: Pimpernel Press, 2020), 19.

19 David E. Cooper, 'Gardens and the Way of Things', in *Earth Perfect: Nature, Utopia and the Garden*, ed. A. Giesecke and N. Jacobs (London: Black Dog, 2012), 20–33.

20 Harrison, *Gardens: An Essay on the Human Condition*, 74.

21 Maggie Keswick, *The Chinese Garden: History, Art and Architecture* (London: Academy Editions, 1978), 82.

22 Yuriko Saito, 'Japanese Gardens: The Art of Improving Nature', *Chanoyou Quarterly*, 83 (1996): 59.

23 A. C. Graham, *Disputers of the Tao: Philosophical Argument in Ancient China* (La Salle, IL: Open Court, 1989), 14.

24 Martin Heidegger, *The Question Concerning Technology, and Other Essays*, 1st ed. (New York: Harper & Row, 1977), 23ff.

25 Kidd, 'Gardens of Refuge', this volume.

26 B. Ziporyn, ed. and trans., *Zhuangzi: The Essential Writings* (Indianapolis, IN: Hackett, 2009), Ch. 12.

27 Mary Midgley, *Beast and Man: The Roots of Human Nature* (New York: New American Library, 1980), 18–19.

28 See Cooper, *A Philosophy of Gardens* (Oxford: Oxford University Press, 2006), Chs 7–8, and Cooper, 'Gardens and the Way of Things'.

7

Gardens of refuge, innocence and toil

Ian James Kidd

The world and I shall have nothing more to do with one another.
TAO YUANMING (365–427)

Introduction

A rhetoric of refuge and escape is a consistent feature of the world's great garden traditions. Harold Nicolson – co-creator, with Vita Sackville-West, of Sissinghurst – called this garden 'a succession of privacies', offering a 'series of escapes from the world'.[1] What one wants to escape from, and how certain gardens enable it, are my questions in this chapter. While my focus is on the Chinese tradition, the aspirations of escape and refuge are enduring features of many garden cultures. I once had a green-thumbed neighbour, taciturn but prone to aphorism, who explained to me that he finds in his garden what cannot 'be reliably found any more in the world'. He mentioned calmness, care and wholesomeness. On another occasion, my neighbour said his garden was a place to 'remember' how life 'used to be' for us and, perhaps, 'how life should be'.

The connections between a desire for escape, need for refuge and disquieting sense that life is no longer what it ought to be gestures to a complex conception of garden appreciation. I explore these connections using Christian, Islamic and Chinese garden traditions. In them one finds a conception of certain gardens as places of *moral refuge* from the corruption and failings of the mainstream world. Within the garden, one enjoys a sense of relief from that world – a world judged to have lost its way, as Daoists might say. Within garden refuges one might be able to recover from the travails of that world. In some of the traditions I discuss, one may also experience some retrieval of a lost kind of moral innocence.

I begin with the idea of *gardens of refuge*.

Refuge

Gardens are designed and appreciated for very many kinds of reasons: places to relax, exercise, be creative, cultivate new life, 'commune with nature', study, meditate or enjoy quiet time away from the world. Not all of these functions involve complex philosophical claims about human moral innocence or a misanthropic need for refuge. However, the deeper conceptions of gardens – as places of refuge – must be taken seriously. For one thing, a rhetoric of escape and refuge is still a feature of our garden-discourse. One might 'escape' into the garden – to water, prune, or sit with G&T – to cool down after a stressful day. However, there are other, less hedonistic senses of escape. David E. Cooper in an interesting paper describes a type of 'sanctuary garden', offering 'a place of retreat or escape from the world outside'.[2] Sanctuaries and refuges are hoped to be safe and secure, in contrast to dangers and threats posed by one's current situation. The filmmaker Derek Jarman used his garden at Prospect Cottage as a refuge from illness, grief, and a hostile news media.[3] Others find their gardens to be refuges from discouraging and oppressive aspects of life at it has come to be – a world of noisiness, ugliness, stress and demands. Many modern champions of 'slow' ways of life list gardening among the practices that help one to resist the 'cult of speed' overtaking modern life, alongside yoga, painting, making music and *qigong*.[4]

The objectionable aspects of the world which inspire a desire for refuge can include *moral harms*. By that term, I refer to attitudes, habits and ambitions that, if internalized, jeopardize one's capacity to live a good life. Moral harms, in this sense, include restlessness, envy, greediness, selfishness, mindlessness and destructiveness. A restless person is perpetually agitated and dissatisfied, incapable of the cool quietude of someone at home in the world. A mindlessly destructive person harms plants, trees, birds and insects – all failures of sensitivity, compassion and wisdom. In the rhetoric of gardens of refuge, such moral harms are judged to be endemic to the mainstream human world. In its goals and ambitions and ethos, the critic sees entrenched failings and their consequences. Our ruthless, fractious world is fuelled by what the Buddha called the 'unwholesome roots' of greed, hatred and delusion. The outcome are social and environmental catastrophes which harm humans, animals and the natural world.

In *Gardens Through History*, William Adams says that the garden, once 'a place for man to escape from the threats of nature', afterwards became 'a refuge from men'.[5] Robert Pogue Harrison opens his 'essay on the human condition' by describing the garden as a 'sanctuary' from the 'rage, death, and endless suffering' of human life. A garden offers 'a kind of haven, if not a kind of heaven'.[6] My misanthropic neighbour would agree. In these testimonies, gardens – some kinds, anyway – are refuges from the moral harms endemic to modern life. It is this – not mere noise or turmoil – from which the wise seek refuge.

Gardens of refuge have three related functions. To start with, they offer *respite*, restorative relief from the battering pressures, distractions and demands of mainstream life. In Roman gardens, those weary of public life enjoyed gardens as places of leisure and relaxation.[7] A *villeggiatura* offered calm, peace and quietude, of a sort impossible in the city. In Pliny the Younger's villa in Tuscany, 'there is no need for a toga, the neighbours do not come to call, it is always quiet and peaceful'. Palladio, likewise, spoke of gardens as refuges for 'the spirit tired of the turmoil of the city'.[8] In classical Greece, the community of Epicurus famously dwelt in the Garden, a perfect environment for a life of equanimity, in contrast to the oppressive 'prison-house' of city life.[9] In China, similar themes of 'respite' echo in its garden writing. Ji Cheng (1582–*c.* 1642) appreciated gardens that offer 'a pure atmosphere'.[10] Chen Fuyao (b. *c.* 1612), a Qing dynasty scholar, enjoyed gardens that offered 'protection against everything . . . most detrimental to the spirit'.[11]

Such aspirations to self-seclusion are not symptoms of antisocial grumpiness. In classical Greece the desire to be alone was often pathologized, since most people regarded the energetic ethos of city life as the ideal *modus vivendi*.[12] All the Roman, Greek and Chinese garden-lovers just cited, however, could endorse the ideals of their societies – but still come to feel that sustained exposure to those societies would erode their virtues. Moreover, their need for respite was justified. Pliny the Younger, Tao Yuanming and Chen Fuyao all felt moral discontent due to long exposure to the corrupting pressures of the world.

The experience of respite from further damage matters but may not help with the damage done. A second function of gardens of refuge is therefore to enable *restoration*. The respite makes possible forms of moral repair – for 'care of the self'.[13] Our damaged moral virtues and crumpled sensibilities, finally protected from ongoing damage, can begin to heal. These days, this ideal of moral recovery is prone to be occluded by focus on the physical and mental health benefits of gardens.[14] But there are moral senses of health and recovery, too. In his novel *Elective Affinities*, Goethe spoke of gardening as giving to life an 'unruffled consistency', as one does 'precisely what needs to be done', and nothing more.[15] Wittgenstein, during a period of mental anguish, worked as a monastery gardener, which fulfilled a need for 'regular work' – not paid work, but a meaningful occupation able to give his emotionally fractious life the constancy and stability it lacked.[16] In the Chinese tradition, a favourite metaphor for moral restoration is washing off the 'dust and grime' of the world. Others associate restoration with awakening or liberation. The poet Tao Yuanming was able, in his garden, to go 'back to nature' – a place for 'waking up' to the true 'meaning of life'.[17] Sima Guang (1019–86), the historian and writer, felt that in his garden, life was at last 'under [his] own control, alone and uninhibited'.[18] Restoration, as a moral attainment, is here associated with being cleansed, awoken and freed – no longer contaminated, insensible or 'inhibited'.

A third function of gardens of refuge is to afford spaces for *reflection* on one's life. Such reflectiveness is a crucial aspect of a good life for the European and Chinese figures cited so far. It also requires receptive conditions, ones that often characterize gardens. Ideally, reflectiveness requires an environment which is still – but not static; attractive – but not distractingly so; and constantly affording new inspiration without being overstimulating. Cooper captures this delicate balance using Rousseau's conception of *reverie*. This is not idle daydreaming. Reverie combines perception, memory, reflectiveness and expansive imaginativeness. In a reverie – like those richly described in Rousseau's 'solitary walks' – our minds flow in unforced, spontaneous ways.[19] Many environments are receptive to reverie, including some artificial ones, but the garden, argues Cooper, offers optimal 'conditions conducive to *reverie*'.[20] Gardens are familiar, calm, secure, attractive, and typically free from distractions and intrusions. In a garden, something is always happening to stir memories, prompt thoughts and sustain imaginings.

Reflection relates to gardens of refuge in two ways. As some recent writers have reminded us, one can take refuge in the mind: in thoughts, ideas, philosophy and contemplation.[21] Not all refuges are places, let alone green ones. Gardens, as spaces for calm, repose and *reverie*, can be spaces for reflection. A second connection is related to the themes of respite and restoration. Once I am undistracted, cleansed and able to 'wake up' – perhaps for the first time in a long time – I am finally able to *reflect* on my life. The best-known expression of this hope is the best-known of Andrew Marvell's poems on gardens, first published posthumously in 1681:

> Fair Quiet, have I found thee here,
> And Innocence, thy sister dear?
> Mistaken long, I sought you then
> In busy companies of men.
> Your sacred plants, if here below,
> Only among the plants will grow;
> Society is all but rude
> To this delicious solitude.[22]

I suggest that the moral functions of gardens of refuge are to provide *respite* and enable *restoration* and *reflection*. In these gardens one finds refuge from the world's aesthetic, moral and existential ills. One can recover and, at least for a time, as Tao Yuanming put it, 'wake up' to the 'true meaning of life'. His poem *Home Again!* (*Gui qu lai ci*) is credited with establishing, in Chinese literary history, 'the ideal of the home and garden as a personal

PLATE 1 Louis-Jacques Mandé Daguerre, *Diorama*, 1842, oil on canvas, 5.4 × 6 meters. Église de Saint-Gervais et Saint-Protais de Bry-sur-Marne. Photo credit © Mathieu Lombard/Musée Adrien Mentienne, Ville de Bry-sur-Marne.

PLATE 2 Stowe Park, Buckinghamshire, The Temple of Ancient Virtue. Photo credit © Martin Pettitt.

PLATE 3 Du Jin, *Tao Yuanming enjoying chrysanthemums*, late fifteenth or early sixteenth century. Hanging scrolls; ink and colour on paper. The Metropolitan Museum. Image in public domain.

PLATE 4 Hermes Courtyard, Plaz Metaxu. Photo: Neil Hepworth Photography / RHS.

PLATE 5 Artemis, Plaz Metaxu. Photo: Author.

PLATE 6 Rilke Inscription, Plaz Metaxu. Photo: Author.

PLATE 7 View over Orexis, Plaz Metaxu. Photo: Author.

PLATE 8 William Kent, *Alexander Pope's garden*, *c.* 1725. © The Trustees of the British Museum.

PLATE 9 Dolmen circles in 1990, front garden, Prospect Cottage. Photo: Author.

PLATE 10 Derek Jarman, *Dungeness Landscape*, 1992, Keith Collins Will Trust. Photo: Author.

retreat, haven, paradise, and world apart'.[23] Such moral functions are consistent with accounts of the more prosaic benefits of gardens and gardening. As Susan Bowden-Pickstock emphasizes, our activities in the garden can, if done properly, be 'beneficial in a physical, mental, emotional, and even a spiritual sense'.[24] In Chinese and Japanese traditions, those dimensions of life are understood to be interdependent. For a flourishing human being, 'the aesthetic and the ethical are both bound up with the cultivation of mental and physical well-being'.[25]

In a garden of refuge, one is protected from whatever erodes our capacities for virtue, peace and flourishing. Moreover, one is able to recover and engage in reflection or *reverie*. If so, a harried person, damaged by the moral harms of the world, could desire to find or create their own garden of refuge. I worry, though, that, left at this, the account is too positive. Unless amended, it omits an important, if darker aspect of gardens of refuge.

To introduce this other aspect, I turn to the notion of *innocence*.

Innocence

In the famous Genesis account of creation, Adam and Eve were in a state of innocence. Satan tricked them into eating from the Tree of Knowledge of Good and Evil, which destroyed their innocence. All this took place in a garden. Christian theology offers rich analyses of the nature of innocence and our potential for salvation. Without engaging in the intricacies of those reflections, I suggest that *innocence* is an apt term for reflection on the *limits* of gardens of refuge.

I suggested that a good garden of refuge, by virtue of its quietude and ambience, can offer respite from the hazards of the world and recovery from damage one suffers. While this may be right, there are two important limitations. The first is that respite and rejuvenation, in the relevant senses, are not always possible. Some pressures are too powerful, the noise of the world too overwhelming. Certain kinds of damage are also too severe to admit of recovery. This may be a pessimism in Wordsworth's famous lament:

> The world is too much with us; late and soon,
> Getting and spending, we lay waste our powers

If our powers are 'wasted' to the point of atrophy, their restoration may be impossible. The Daoist philosopher Zhuangzi (*c.* 369–*c.* 286 BCE) exclaimed 'How sad!' when seeing people 'entangled' by the kinds of artificiality that 'tie down Virtuosity'.[26] The *Daodejing*, too, saw such people as having lost the Way, casting them into greed and violence.[27] Wise people, said Zhang Heng (78–139), realize the Way is 'benighted', so retreat into their gardens to avoid the 'dust and grime' unavoidable in a life devoted to 'worldly affairs'.[28]

A problem with metaphors of being 'entangled', 'tied down' and 'dust and grime' is that they imply the possibilities of being freed, released, or cleansed. Of course, these refer to respite and restoration. A different set of metaphors are less encouraging. When the Buddha spoke of the world as 'burning' with the 'fires' of hatred and greed, he points to more permanent kinds of damage, as do his other images, like a leper frustratedly clawing at their sores. So, even in a near-perfect garden of refuge, one may find little respite and no recovery in the desired sense. The world, with its pressures and ugly realities, is 'too much with us', so one can never, in the fullest sense, experience 'Fair Quiet'.

A second limitation, this time of the rejuvenation and reflection functions of gardens of refuge, refers us back to the initial desire to seek sanctuary. The reflections and the spontaneous *reveries* in a garden could take many forms: nostalgic reminisces, idle wonderings, warm hopes for the future, disinterested absorption in the busyness of birds. A person disappears into what Gaston Bachelard called 'the space of elsewhere' to gently follow its 'reverberations'.[29] Sometimes, however, *reveries*, even if following their own course, will lead us back to the dark vision of the world. Perhaps I find myself immersed, content in my garden, admiring the delicately determined flight of a butterfly. I wonder what its world is like, and if it is as unworried as it seems – then am reminded of all the worries of my own life. Or during a *reverie* I reflect on the pleasure of 'delicious solitude', and idly wonder what experiences prompted Marvell himself to speak of 'busy companies of men' whose life is 'all but rude' – then reflect on grim experiences in my own life. In these cases, the reflections a refuge can enable naturally include, among other topics, the moral harms that led one to desire 'escape' in the first place.

The two limitations are closely connected. Aspirations to seek refuge are continuous with a sense of the harms or hazards of the human world. What one enjoys in a garden of refuge – contended calm, stillness, rejuvenation, peace, 'delicious solitude' – can also be experienced as absent from one's wider life. Moreover, their absence is understood not as a product of one's bad luck, but as a consequence of the demandingness and relentlessness of life. Put another way: what leads us into a garden of refuge can also bring one out of it, at least psychologically. If so, we can restate the connection between gardens of refuge and innocence. The moral harms of the world include an erosion of our moral innocence. In those gardens, one is reminded of modes of comportment uncorrupted by the failings of the world.

The theme of lost innocence is there in many classic accounts of gardens of refuge. Tao Yuanming's fable of a man who stumbles into a hidden community of contented, peaceful people – 'Peach Blossom Spring' – has been said to voice 'a universal longing for innocence and happiness'.[30] The world is hidden, though, and its distinctive moral attractions seem to depend on its isolation – and difference – from wider society. By virtue of that isolation, the inhabitants are able to 'preserve their purity and innocence'.[31]

In the Garden of Epicurus, one can cultivate equanimity and virtue in ways impossible to most Athenians. Within the Garden, there are, by design, few inducements to selfishness, greed or hatred. Recall, for instance, the deliberate simplicity of the life of the Epicureans, for whom a 'pot of cheese' counted as a luxury. The virtuous life made possible in the Garden, then, entailed disciplined resistance to temptations that a morally innocent person would not experience, nor have to struggle against.[32]

Lost innocence is a better metaphor for communicating this sense of lasting moral damage. Innocence is a quality that, once lost, cannot be regained. It can be destroyed by awful experiences or it will naturally fade away, as a child is initiated into the complexities of the world. Innocents like children spontaneously act kindly and honestly because it does not yet occur to them to do otherwise. It is temptation and the capacity for calculating ways of thinking that feed vices such as dishonesty and manipulativeness. This is the reason that innocent modes of moral comportment, as long as they last, are attractive. Innocent people are not yet subject to worldly temptations and pressures, effective resistance to which demands self-discipline and self-restraint.

I suggest that gardens of refuge can induce a sense of one's lost moral innocence. When my neighbour spoke of remembering 'how life used to be', he referred, I suspect, to a kind of moral comportment now all but impossible except in very specific contexts and environments. Contented immersion in ways of engaging with the world in gentle, thoughtful, unselfish ways are possible in a 'refuge' – but not beyond its walls. In Genesis, the Garden of Eden remained peaceful, even as the world beyond its borders became 'cursed' and a place of 'thorns and thistles' and, for humans, suffering, alienation and 'painful toil'.[33] In gardens of refuge, one has fleeting and fugitive intimations of that lost innocence.

The theme of innocence in relation to gardens of refuge should recall the 'paradise gardens', popular in the Islamic and Christian traditions of the seventh to the thirteenth centuries. The ideal paradise garden is lush, fertile and well-watered, as well as calm, orderly and peaceful. It symbolizes both nature and the human relationship to it – as originally intended by God.[34] *Paridaeza* – an old Persian word – meant 'walled garden', a favourite Koranic image of Heaven.[35] Life should have been hospitable and free of intruding moral discontent and any fear of corruption. Paradise gardens evoke a time when, as Milton puts it in *Paradise Lost*, living was 'a delightful task'.[36] Of course, these gardens also serve a second, darker purpose for Muslims and Christians. Eden, the site of our original sin, 'reminds us of temptation, transgression, and loss'.[37] Paradise gardens, by recalling our Edenic state, can also help us 'anticipate heaven, to which they should aspire'.[38] Experiences of paradise gardens are shaped by a postlapsarian image of humankind as corrupted – 'fallen' beings incapable of true happiness in worldly life, albeit destined to enjoy it after death. As Carolyn Merchant neatly summarizes the story:

> The Christian story is marked by a precipitous fall from a pristine past. The initial lapsarian moment, or loss of innocence, is the decline from garden to desert as the first couple is cast from the light of an ordered paradise into a dark, disorderly wasteland to labour in the earth.[39]

Here is a specific conception of a garden of refuge, shaped by monotheistic religion. It connects gardens to 'retreat, contemplation, and repose' but with a 'darker resonance' owing to the themes of 'fallenness', lost innocence and moral disaster.[40]

Paradise gardens are tied to the doctrines and narratives of Christianity and Islam – or, better, to specific variations within those rich religious traditions. Not all Christians see the Fall as the annihilation of our moral innocence. St Aquinas was less pessimist about this than St Augustine, for instance. However, the rich connections between gardens of refuge, moral innocence and human corruption are not unique to Islamic and Christian discourses. Similar themes are also explored further east in Chinese garden traditions.

Toil

The schools of the classical period of Chinese philosophy, for all their differences, shared a profound sense of moral and cultural decline. The nature and causes of decline, and the best means of response, were matters of debate. The 'Agriculturalist School' (*Nongjia*), one of the Hundred Schools of the classical period of Chinese philosophy, told a story of early and simple ways of life – small communities of people, bound together by trust and mutual acquaintance. Life was less complex. Desires were modest and attainable by hard work mainly done by shared labour. There was no war, complex governance or hubristic confidence in human powers. Life proceeded without 'rewards or punishments, commands or restrictions', for it was recognized that these are 'a means to disorder'. Over time, this way of life deteriorated. Men became 'shameless and careless', an appetite for 'gain' corrupted their hearts, until disorder became 'inevitable'.[41]

In the texts gathered under the label 'Daoism', one finds similar themes. The anonymous authors of the poetic classic, the *Daodejing*, and the book named for Zhuangzi lament our decline from an earlier condition of simplicity and moral innocence. The unknown authors of the *Daodejing* describe a proliferation of rules, desires, regulations, appetites and ambitions that corrupt people, who became 'cunning', hypocritical, arrogant and exploitative.[42] The Inner Chapters of the book of Zhuangzi – the ones most likely authored by him – lament our inflexibility and 'calculating', rigid ways of thinking and our inflexibility.[43] Instead of humble and attentive engagements with people and animals, most people prefer to 'impose' and 'contend' on them, including even well-meaning people, who inevitably 'end up tearing the world to pieces'.[44]

The Agriculturalist and Daoist schools agree on many general details of this narrative of decline, even if they are marked by considerable disagreement and diversity. Some schools of the classical period offer their own narratives of decline, but of a different sort. Confucians and Mohists, while agreeing on the lamentable state of the world, come to different conclusions. Confucians lamented the deterioration of rituals and traditional cultural arts, whereas the Mohists criticized the warlikeness and elitism of Chinese society and the suffering it caused the common people. Others, such as Han Feizi (*c.*280–233 BCE), rejected the preoccupation with 'the ways of the ancients', since the problems of the present differ too much from those of the past.[45] So, while most of the Chinese schools experienced their world as being in moral crisis, not all are keen to connect that to a sense of decline from an earlier period of innocence.

The Agriculturalist and Daoist narratives of decline are related to gardens, however, in instructive ways. What one finds in those texts are appreciations of the edifying power of natural environments lacking or less prominent in the other schools. The *Nongjia* – named for Shennong, the 'divine farmer' – were pastoralists who caringly worked the land. In the 'Daoist' case, this is truer of the *Zhuangzi* than the *Daodejing* – the latter too terse anyway to dictate definitive interpretations. But into the later Daoist philosophical and religious tradition there is a marked appreciation of the importance, to a good life, of engagements with natural places and creatures.[46] Some admirers have described Daoists as 'gardeners of the cosmos', a figurative term meant to capture their commitment to modes of relating to natural creatures and places marked by skill, sensitivity and understanding.[47] Such people aspire to 'slowly shape their life and environment', appreciatively alert to an intimacy of 'human activity and the order of nature'.[48] While this is the ideal, it has become increasingly difficult for humans, who become ever more complex and artificial. In a sense, there has been a shift from our being *gardeners* to *technologists* – to hubristic beings who 'set on' and exploit what we would once have nurtured and cultivated.[49]

The sense of innocence, decline and appreciation of natural environments, including 'humanized' ones, such as gardens, echoes throughout the later Chinese tradition. As historians of Chinese literature emphasize, Daoist writings exerted enormous influence on later poetry and landscape art.[50] Much of this took place in gardens of refuge. Ji Cheng wrote in the *Yuan ye* (*The Craft of Gardens*) of 1631 of his desire for a well-constructed garden, offering 'a pure atmosphere around our tables and seats [so that] the common dust of the world is far from our souls'.[51] Sima Guang created his 'Garden of Solitary Delight' in which, for a time at least, he could live 'uninhibited'.[52] The scholar-recluse, Zheng Yuanxun (1603–44), created his 'Garden of Reflections' as a retreat where he could devote himself to care of his elderly mother and study of his books. Invoking the respite theme, Zheng contrasted the caring and gentle activity of 'watering the trees and the flowers' with febrile worldly anxieties about 'establishing merit and making a reputation'.[53] In these and other testimonies, a garden of refuge offers an environment which enables and recalls more innocent kinds of moral comportment.

In some Chinese poetry these experiences are tinged by moods of resignation, nostalgia and lament. One feels this in poetry of Wang Wei (699–759):

> In old age I ask for peace
> and don't care about things of this world
> I've found no good way to live
> and brood about getting lost in my old forests.
> The wind blowing in the pines loosens my belt,
> the mountain moon is my lamp while I tinkle
> my lute.[54]

Gardens of refuge, when reflected on, can induce a series of moods. These include a poignant sense of our lost moral innocence, a longing nostalgia for former ways of being-in-nature, and a misanthropic sense of the moral dangers of our life. One could express this in a sense of the *moral toil* – the demanding moral work – of modern life. In an overactive world of constraints, demands, trade-offs, temptations and pressures it becomes ever harder to cultivate, exercise and protect our virtues. Life becomes *harder* in ways that were not features of our innocent life. As Adam and Eve found, existence outside Eden is hard work – a point that can be detached from its specific theological articulation. *Moral toil* implies kinds of moral comportment that are difficult, stressful and often unrewarding. It conveys a sense of our moral life as hard, imperilled, far from the 'delightful task' that, for Milton, it used to be. In a garden, at least for a time, it becomes possible for a reflective person to experience and appreciate those more innocent forms of moral life.

A sceptic might reply that these claims about gardens, refuge and moral innocence are too 'high-falutin', a pretence that our time in a garden is a matter of existential drama. Such scepticism is typical of those unaware or unable to realize the moral and spiritual resonances of gardens across many of the world's cultures. Several hours spent perusing work on the philosophy and theology of gardens should correct that obliviousness. In any case, there are many testimonies to these ways of experiencing and understanding gardens, including among contemporary gardeners. Other than my neighbour, the American-Canadian author, Ruth Ozeki, in her book *All Over Creation* has a character – a Japanese-American woman – whose heart 'crack[s] open' at all the 'hopeless beauty and fragility and loss of all that is precious on earth'. This experience occurs in a Californian greenhouse – air-conditioned, artificial – which recalls and contrasts with her Hawai'ian childhood home to which she cannot return. Her youth – unhurried and unworried, calm and happy – was a place and a time when 'everything is a garden'.[55]

Appreciation of gardens of refuge incorporates a sense that they enable kinds of moral comportment which recall, however dimly, a more innocent life. In these gardens, the corrupting imperatives of the world are absent, enabling a welcome experience of respite and rejuvenation. In certain moments, however, one may be led into reflections and *reveries*, and from there back into reflection on the moral harms and toil of the world beyond the garden wall.

Conclusions

In a remark widely quoted in the philosophy of gardens, Yuriko Saito once said a Zen garden offers a 'glimpse of the world as it appears to a Zen-enlightened sensibility'.[56] 'Glimpse' is the right word – a brief, fleeting sighting of something passing by, or perhaps passing away. The Southern Chan School of Buddhism, initiated by Huineng (638–713), emphasized gentleness, spontaneity, humility, and these passed into the later Zen tradition of Japan.[57] While in practice this was physically demanding, requiring perseverance and self-discipline, it was informed by a sense of innocence, spontaneity and gentle responsiveness.[58] I think this is what my neighbour meant when explaining that he finds in his garden what he can no longer 'reliably find' in the world: echoes of morally innocent life characterized by 'quieter' virtues – peacefulness, contentedness, stillness, quietude, gentleness, loving care, and an unforced sense of *intimacy* between the human and the natural. These innocent ways may once have prevailed in the world, but have now, alas, become so difficult, in a world of toil, that they could only really be 'remembered'. Such realization brings with it an abiding set of moods – poignant, melancholic, pessimistic, even misanthropic.[59] Refuges are made or sought to protect us against a hostile world, one of toil and corruption. Sometimes, however, a refuge cannot in the end protect itself against that world.

Acknowledgements

I am grateful to the editors, especially to Yue Zhuang, for inviting me to speak at the Exeter workshop from which this volume originated, to the speakers and attendees at that event, and to David E. Cooper for comments on a draft.

Notes

1 Quoted in Tim Richardson, *Sissinghurst: The Dream Garden* (London: The Quarto Group, 2020), 143.

2 David Cooper, 'Gardens, Nature, and Culture', *Journal of Scottish Thought* 9 (2017): 1–13 (7).

3 Michael Charlesworth, '"Each stone has a life to tell": Derek Jarman's garden at Prospect Cottage', this volume.

4 Carl Honoré, *In Praise of Slow: Challenging the Cult of Speed* (London: HarperCollins, 2005).

5 William H. Adams, *Gardens through History: Nature Perfected* (New York: Abbeville, 1991), 319.

6 Robert Pogue Harrison, *Gardens: An Essay on the Human Condition* (Chicago: University of Chicago Press, 2010), ii–iii.

7 Katharine T. von Stackelberg, *The Roman Garden: Space, Sense, and Society* (London: Routledge 2009), 96.

8 James S. Ackerman, *The Villa: Form and Ideology of Country Houses* (Princeton: Princeton University Press, 2023), 13.

9 Meghan T. Ray, 'Cultivating the Soul: The Ethics of Gardening in Ancient Greece and Rome', in *Gardening – Philosophy for Everyone: Cultivating Wisdom*, ed. Dan O'Brien (Oxford: Blackwell, 2010), 220–31.

10 Ji Cheng, *The Craft of Gardens*, trans. A. Hardie (New Haven: Yale University Press, 1988), 124.

11 Quoted in Cooper, 'Gardens, Nature, and Culture', 8.

12 Rosalia Hatzilambrou, 'Being Unsociable in Classical Athens: A Very Bad Attitude Indeed', in *Being Alone in Antiquity: Greco-Roman Ideas and Experiences of Misanthropy, Isolation and Solitude*, ed. Rafal Matuszewski (Berlin: Walter de Gruyter, 2021), 149–66.

13 David E. Cooper, *A Philosophy of Gardens* (Oxford: Oxford University Press, 2006), 86ff.

14 Isis Brook, 'Gardening and the Power of Engagement with Nature for Mental Well-being', in *The Oxford Handbook of Mental Health and Contemporary Western Aesthetics*, ed. Martin Poltrum, Michael Musalek, Helena Fox, Kathleen Galvin, Yuriko Saito (online edn, Oxford Academic, 16 August 2023), https://doi.org/10.1093/oxfordhb/9780192866929.001.0001, accessed 8 September 2024.

15 J. W. von Goethe, *Elective Affinities*, trans. Richard Hollingdale (Harmondsworth: Penguin, 2017), 24.

16 Quoted in Cooper, *A Philosophy of Gardens*, 7.

17 Quoted in Joseph Cho Wang, *The Chinese Garden* (Hong Kong: Oxford University Press, 1998), 58.

18 Quoted in Ji Cheng, *The Craft of Gardens*, 124.

19 Jean-Jacques Rousseau, *The Reveries of a Solitary Walker*, trans. C. Butterworth (Indianapolis: Hackett, 1992).

20 Cooper, *A Philosophy of Gardens*, 76.

21 Zena Hitz, *Lost in Thought: The Hidden Pleasures of an Intellectual Life* (Princeton: Princeton University Press, 2010), ch. 1.

22 Quoted in Thomas D. Church, *Gardens are for People* (Berkeley: University of California Press, 1995), 6.

23 Richard M. Barnhart, *Peach Blossom Spring: Gardens and Flowers in Chinese Paintings* (New York: Metropolitan Museum of Art, 1983), 15.

24 Susan Bowden-Pickstock, *Quiet Gardens: The Roots of Faith?* (London: Continuum, 2009), 123.

25 Graham Parkes, 'The Role of Stone in the Chinese Rock Garden', in *The Bloomsbury Research Handbook of Chinese Aesthetics and Philosophy of Art*, ed. Marcello Ghilardi and Hans-Georg Moeller (London: Bloomsbury, 2021), 277–300 (277).

26 *Zhuangzi: The Essential Writings with Selections from Traditional Commentaries*, translated by Brook Ziporyn (Indianapolis, IN: Hackett, 2009), ch. 23.

27 *The Daodejing of Laozi*, trans. Philip J. Ivanhoe (Indianapolis: Hackett, 2002), ch. 53.

28 Quoted in Timothy Wai Keung Chan, *Considering the End: Mortality in Early Medieval Chinese Poetic Representation* (Leiden: Brill, 2012), 193.

29 Gaston Bachelard, *The Poetics of Space*, trans. M. Jolas (Boston: Beacon, 1994), 184, xv.

30 Sunpyo Hong and Chin-sung Chang, 'Peace Under Heaven: Confucianism and Painting in Early Joseon Korea', in *Art of the Korean Renaissance, 1400–1600*, ed. Soyoung Lee (New Haven: Yale University Press, 2009), 65–92 (75).

31 Xingpei Yuan and David R. Knechtges, 'Tao Yuanming 陶渊明 (365–427)', in *Ancient and Early Medieval Chinese Literature: A Reference Guide*, ed. David R. Knechtges and Taiping Chang (Leiden: Brill, 2013), 1090–124 (1111).

32 Gordon Campbell, 'Epicurus, the Garden, and the Golden Age', in *Gardening – Philosophy for Everyone: Cultivating Wisdom*, ed. Dan O' Brien (Oxford: Blackwell, 2010), 220–31.

33 Genesis 3: 17-18.

34 Nerina Rustomji and Nari Rustomji, *The Garden and the Fire: Heaven and Hell in Islamic Culture* (New York: Columbia University Press, 2009).

35 Quran 44:52 and 47:15.

36 John Milton, *Paradise Lost, a Poem in Twelve Books* (Glasgow: Printed and sold by R. & A. Foulis, 1672), vol. 4, 437.

37 Stephen Bending. *Green Retreats: Women, Gardens and Eighteenth-Century Culture* (Cambridge: Cambridge University Press, 2013), 216.

38 Cooper, 'Gardens, Nature, and Culture', 2.

39 Carolyn Merchant, *Reinventing Eden: The Fate of Nature in Western Culture*, 2nd ed. (London: Routledge, 2003), 12.

40 Stephanie Ross, *What Gardens Mean* (Chicago: University of Chicago Press, 2001), 6.

41 Quoted in A. C. Graham, *Disputers of the Tao: Philosophical Argument in Ancient China* (Chicago: Open Court Publishing Co., 1989), 67.

42 *Daodejing*, chs 13, 18, 75.

43 *Zhuangzi*, chs 1, 4, 6.

44 *Zhuangzi*, ch. 33.

45 *Han Fei Tzu: Basic Writings*, trans. Burton Watson (New York: Columbia University Press, 1964), 340.

46 David E. Cooper, *Convergence with Nature: A Daoist Perspective* (Dartington: Green Books, 2012); James Miller, *China's Green Religion: Daoism and the Quest for a Sustainable Future* (New York: Columbia University Press, 2017).

47 David E. Cooper, '"Gardeners of the Cosmos": The Way of the Garden in East Asian Tradition', in *The Good Gardener: Nature, Humanity and the Garden*, ed. Annette Giesecke and Naomi Jacobs (London: Artifice, 2015), 38–49.

48 James Miller, *Daoism: A Short Introduction* (Oxford: Oneworld, 2003), 45–6.

49 Martin Heidegger, *The Question Concerning Technology and Other Essays*, trans. William Lovitt (New York: Harper & Row, 1977).

50 Stephen Little and Shawn Eichman, *Taoism and the Arts of China* (Chicago: The Art Institute of Chicago, 2000); Chung-yuan Chang, *Creativity and Taoism: A Study of Chinese Philosophy, Art and Poetry* (London and Philadelphia: Singing Dragon, 2011).

51 Ji Cheng, *The Craft of Gardens*, 44.

52 Quoted in Ji Cheng, *The Craft of Gardens*, 124.

53 Zheng Yuanxun, 'A Personal Record of My Garden of Reflections', trans. and intro. Duncan M. Campbell, *Studies in the History of Gardens & Designed Landscapes*, 29, no. 4 (2009): 270–81 (276–7).

54 Wang Wei, *Laughing Lost in the Mountains: The Poetry of Wang Wei*, trans. Tony Barnstone, Willis Barnstone and Xu Haixin (Hanover, NH: University Press of New England, 1991), 3.

55 Ruth Ozeki, *All Over Creation* (Edinburgh: Canongate, 2022), 454, 450.

56 Yuriko Saito, 'Japanese Gardens: The Art of Improving Nature', *Chanoyou Quarterly*, 83 (1996): 41–61 (59).

57 In Japan, Zen practice became more demanding, requiring immense perseverance and self-discipline from its practitioners.

58 David E. Cooper, 'Garden practices, "quiet" virtues and understanding', this volume.

59 Ian James Kidd, 'Gardens of Refuge', 6 October 2021, https://daily-philosophy.com/kidd-gardens-of-refuge/index.html, accessed 28 August 2024.

8

Withdrawal and bewilderment

Plaz Metaxu, the 'place that is between'

Alasdair Forbes

The things of this world sing out only when in a state of disequilibrium.
HAN YU[1]

I

Why have I chosen to write about my garden in the context of bewilderment? Bewilderment is not among the virtues David Cooper mentions as intrinsic to gardens.[2] Indeed, it is seldom classed as a virtue at all, though there are signs this may be changing, at least insofar as it is now coming to be thought of as a characteristically *poetic* virtue.[3] I am in complete agreement that virtues such as humility, patience and forbearance are necessary and vital to the gardener. I would add to the list gratitude, trust and docility (the latter perhaps the purest expression of what used to be called the *passio essendi*, renamed by the contemporary philosopher William Desmond as the 'porosity of being').[4] I am completely sympathetic to the idea that virtues like these draw attention to our dependence rather than to our autonomy – and that this dependence (usually on 'nature', but also on each other and, perhaps, on more numinous realities called by such names as Buddha Nature, Dao and God) reflects a dimension of our lives that is all too often overlooked, especially in what Lionel Trilling once called an 'iron time' such as our own.[5]

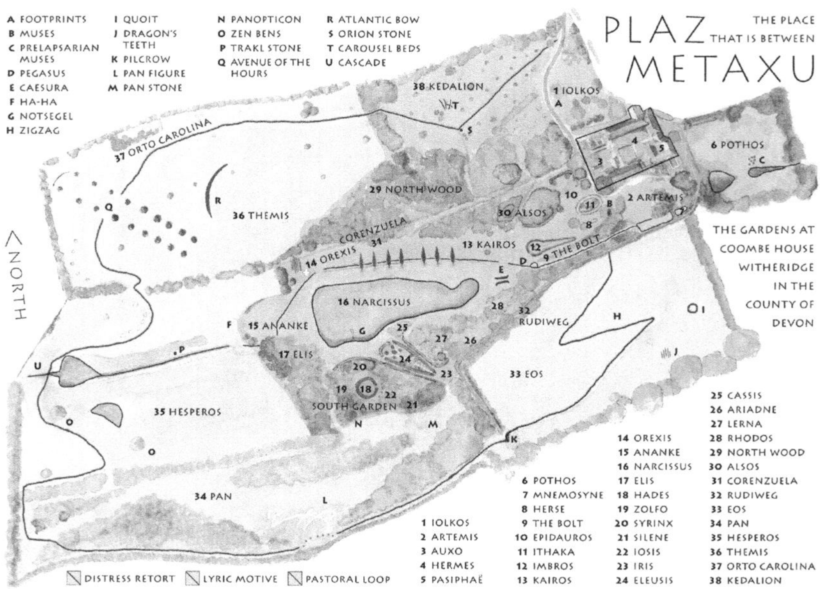

FIGURE 8.1 Garden Plan of Plaz Metaxu. The overall size of the garden is 13 ha (32 acres) including the outlying fields, which are grazed by sheep. A landscaped route through these fields is marked on the plan as the 'Pastoral Loop'. Areas and features mentioned in the text are identified by numbers and letters as they appear on the Plan. Source: Author

'What becomes of us in this state of docility?' is a question I have found gardening makes me ask. I hardly dare speak for others, but, for me, there are further questions, too, that docility gives rise to, which I think deserve mentioning even though this risks upsetting the usual relationship between the arousing and the stilling of desire we favour in gardens, since it suggests that not all peace of mind or contentment is on the same level or of the same sort.[6] For example, is this sense we have of our own dependence always harmonious and reassuring? Do the quiet virtues that foster it always deliver an equable and serene state of mind? Is the sense of unfolding mystery which dependence may detect in the nature of things always composing and never unsettling? How do we become responsible to this mystery and draw it from the margins to the centre of our lives? Such questions entitle us to be at a loss, they allow us to admit that we are sometimes baffled by reality, to recognize that even in tranquillity and beauty it is all right to notice a searching component, an uncanny aura, a rumour of bewilderment. Since gardens, I think, are supremely permissive places, they find room for this bewilderment to flourish alongside our gratitude and our contentment. Bewilderment is perhaps most characteristically manifested in the liminal moment, in which we find ourselves disconcerted or put at risk, but when, too, a possible transformation can occur, which predisposes us to deepen our sense of wonder.[7]

Nature itself is both bewildering and inviting. Gardens celebrate nature's invitation to us to make ourselves at home. As part of this celebration we may be led, through a variety of means, to notice how mysterious this home is, but, just as likely, the contentment we find in a garden disguises the strangeness of our existential situation from us. It seems to me entirely fitting a garden should do either one of these things. But a garden also causes frissons of bewilderment when we experience nature there in its volatility and unpredictability, its prodigality, its uncanny beauty, its decay, its indifference, its violence, its seemingly senseless cycle of reproduction where life feeds remorselessly on life, a fact which so dismayed Leopardi he thought that, on close inspection, a garden was a more depressing place than a cemetery.[8] Vulnerability draws attention to our dependence at least as vividly as trust does. Bewilderment becomes in this sense a way of staying with our vulnerability, just as contentment can be a way of staying with trust. I think, if true equanimity is our goal, it is impossible to dispense with a long education in either.[9]

At a deeper level too, though, nature as the phenomenal world, or in China the 'Ten Thousand Things',[10] has a profoundly mysterious side to it, which is perhaps most comprehensively embraced by those who undertake a religious or poetic withdrawal. Examples would be Yan Hui,[11] as portrayed in ch. 6 of the *Zhuangzi*, where he represents a fictional characterization of the author, and Tao Yuanming[12] and Wordsworth,[13] the archetypal poet recluses in China and England respectively. Yan Hui's example is important, I think, since it indicates that, for the committed Daoist, the reality the Dao represents, especially in its numinous aspect, is only attainable through a long and dedicated inner training (*nei ye*), involving such practices as 'fasting the heart-mind' (*xin zhai*) and 'sitting and forgetting' (*zuo wang*) until 'mind and body fall away' and the individual identity is finally able to merge with the Great Thoroughfare (*da tong*), of which ostensible nature is no more than the phenomenal manifestation.[14] Harold Roth has aptly characterized such practices as 'Daoist apophatic meditation'.[15] Wordsworth, according to Geoffrey Hartman, is also led beyond manifest nature, but this time by nature itself rather than by an inner training, through a process which Hartman likens to a *via naturaliter negativa* (a negative way by means of nature).[16] Whereas Tao Yuanming, when we look beyond the frequently anthologized works to his oeuvre as a whole, oscillates between moods of radiant contentment with his surroundings and melancholic foreboding as to the wider setting of life.[17] One thing I take from these religious or poetic examples of withdrawal is how a sense of the wonder and preciousness of the phenomenal world is heightened rather than repressed when it is made wholly interdependent with an equally vivid experience of that world's mysterious withdrawal, experienced by Wordsworth as 'fallings from us vanishings'[18] and referred to by Daoists as the movement of the Dao back towards the preternatural emptiness. Contact with this 'unknown mode of being'[19] is made involuntarily by Wordsworth, and through what, in the Daoist context, Edward Slingerland calls a *de*cultivation of the self (Roth's 'apophatic meditation').[20] Simone Weil speaks of an imperative towards 'decreation'.[21] For me, any rounded response to what is called nature needs to fully embrace, in Heraclitus' phrase, this love nature has for hiding,[22] and, while for the enlightened sage the hidden and ostensible realms may coexist in a perpetual and mysterious harmony, as exemplified in both Wordsworth and Tao Yuanming the moments of harmony alternate with moments of bewilderment, as is hardly surprising, given (so far as we know) the absence of any formal 'inner training' on the poets' parts, and the disconcerting nature of the reality the world's withdrawal presents to us.

Given this understanding of a nature that is inviting and bewildering in equal measure, the image of a garden I cherish is one that makes a fragile and tactful intervention in its landscape, provisionally poised between the emergence and the withdrawal of the world, as indeed all life is.[23] The gardens that have moved me the most are the ones that seem to me to generate this poise most eloquently, and they do not conform to any single type. Many Chinese and Japanese gardens[24] seem centred on this intuition, for instance, but certain early landscape gardens like those at Courances[25] in France or Studley Royal in Yorkshire[26] also have this effect on me. This is because, or so it seems to me, these gardens both respect and transform their natural settings, investing them with a kind of preternatural poise. In them a rumour from a numinous elsewhere coincides with the gratitude for being here. When I discover such ontological poise, I feel at last that I may know my place. Such gardens seem to confirm, and to exonerate, my interim existence, and that is as much as I can ever ask.

The licence to be bewildered is the part of that exoneration that is least explored in the discourse of gardens, so I hope you will forgive me if I have chosen to write about it here. I scarcely know how helpful or not this may be for the understanding of gardens generally, but I do know that in my own garden bewilderment has played an important role. I will try to describe some of the ways in which bewilderment appears at Plaz Metaxu, first, more self-consciously, in the Hermes courtyard and, second, more fatefully, or so at least it has seemed to me, in the valley at large. The selection of these two faces of the garden should not be considered exclusive: the garden's name as the 'place that is between' gives the best clue that what I have called the 'liminal moment' may in fact occur in the garden anywhere.

The emphasis on bewilderment inevitably gives a rather unbalanced account of the garden. I joked with Yue, my co-convenor of the workshop that has given rise to this book, that it might have been more welcome if I had chosen to write about 'Withdrawal and gratitude'! In fact, no inspiration for making the garden has been more lasting or important than my profound 'gratitude for space', though when I mention this to people (not including

Yue!) it often draws as blank an expression as bewilderment does! When I write about bewilderment or the gods or other things that stir the soul's conscience, it is important to realize they are not supposed to usurp or relegate the spontaneous reactions we have to a garden. The interaction of this spontaneity with our deepest concerns, I would hope to the benefit of both, is nonetheless, in my experience, the very staple of the life of the gardener.

II

The courtyard dedicated to Hermes, the former farmyard, is immediately behind the house (Plate 4; Garden Plan 4).

It is a formidably secluded space, enclosed by stone barns, which I liken to the garden's *ricetto*, or redoubt.[27] As such, it makes a dramatic contrast to the garden in front of the house, which is dedicated to Artemis, the goddess of pristine and virginal nature, where a green gesture of tribute is paid to the surrounding trees, hillside and sky (Plate 5; Garden Plan 2).

At Plaz Metaxu, this sort of counterpoint companionship between different spaces is a fundamental organizational principle of the garden, operating on both a physical and a metaphorical level. The Hermes courtyard is a self-referential and inward-looking space, where many intuitions incubated about what sort of place the garden should be. Protected from the outside world, and sheltered even from the worst of the weather, it was an ideal space in which to explore the virtues and the vicissitudes of withdrawal.

The courtyard's inward-looking space presented itself as a mirror image of the inward turn I needed to make in my mind and heart to fulfil, as I saw it, the opportunities and obligations of withdrawal. There were creative incentives behind my decision to withdraw, which centred on my longing to make a garden. But the mental landscape which necessitated my decision to withdraw was not so much characterized by dissatisfaction with the urban or social worlds I was leaving behind (the classical reason for withdrawal in much of the philosophical literature in both Europe and China) as by a profound bewilderment about the nature of things more generally. I found it less easy to attach blame for any perceived ills of the world to external circumstances than to myself. I knew in in the middle course of my life that 'the direct way was lost' (if it had ever been found) and thus it was that the courtyard and adjacent enclosures became a kind of garden version of Dante's 'dark wood' for me, to which I gave the name of the 'Distress Retort'.[28] (Garden Plan 3, 4, 5). Both the 'dark wood' and the 'distress retort' (which means not only 'antidote to distress' but also the vessel or alembic in which distress can be focused and distilled) imply an initiation by way of bewilderment, with its radical sense of being at a loss, its recursive dismay at suffering and paradox, its openness to unsuspected wonders, its requiring of patience, its broaching of alternatives, its sounding of dread.

I call the courtyard a 'reflexive' space (again in contrast to the front lawn's 'votive' one) and Hermes was the god of that reflex. But mortal actors played their part in guiding and refining that reflex, too. First and foremost, I was deeply impressed by, and indebted to, those contemplative and mystical traditions which, it seemed to me, approached the wonder and the unaccountability of being (including the experience of suffering, evil and paradox) with a candour and depth of commitment, reflected in the gift for a daring and extreme introversion and withdrawal, and a 'purging of the intellect',[29] that is only to be found in those who radically and intimately place every aspect of their lives at the disposal of a transrational and suprahuman power.[30] This species of withdrawal is the one that continues to haunt me most profoundly, its absolute or 'impossible' nature only serving to deepen my respect for it, while highlighting my own sense of bewilderment and inadequacy.[31] For me, the 'inner training' such withdrawal recommends both acts to deepen incomparably our sense of responsibility towards the mysterious ground of being and provides an ultimate sanction for the imperative to withdraw itself. A garden as the setting of withdrawal is nonetheless a supremely permissive place, allowing for many and various expressions of the vocation to withdraw, granting us a licence to experiment with the poetic and the psychological repercussions of what it means to make ourselves a dwelling-place on earth, for instance. A garden's latitude in this respect, its liberal and polymorphous companionship, has been, I should say, an altogether saving influence for a nature such as mine. From among the countless mentors and models who have helped and inspired me in my life of withdrawal as a gardener, I will here single out three representative examples of a philosopher, a poet and a psychologist.

The Spanish philosopher Ortega y Gasset introduced me to the term *ensimimasmiento*, which I might translate as a withdrawal into oneself in search of authentic discernment.[32] As I lack intellectual resilience, and am easily baffled and perplexed by the free-for-all exchange of ideas to be found in the Academy and elsewhere, this philosophical sanction for an introspective withdrawal was important to me as I attempted to work out what really mattered for myself.

For a long time I had loved and admired the work of Emily Dickinson (1830–86). While it may have been true for poets like Tao Yuanming and Wordsworth that – to quote Tao – 'Nothing like the others, even as a child' they were 'rooted in a love for hills and mountains'[33] to which they could make their joyous return when the opportunity arose, I recognized in Emily Dickinson's sense of apartness something that rang truer with my own experience, as when she wrote:

A loss of something ever felt I –
The first that I could recollect
Bereft I was – of what I knew not

Too young that any should suspect
A mourner walked among the children[34]

This augural sense of 'being at a loss' was, however humbly, of pivotal importance for me, too, and as I tried gradually to educate myself, I grafted many other kinds of bewilderment onto it. In a poem such as the following one, I also learned from Dickinson how absolute the focus of poetic withdrawal might need to be:

The Soul selects her own Society –
Then – shuts the Door –
To her divine Majority –
Present no more –

Unmoved – she notes the Chariots – pausing –
At her low Gate –
Unmoved – an Emperor be kneeling
Upon her Mat –

I've known her – from an ample nation –
Choose One –
Then – close the Valves of her attention –
Like Stone – [35]

We might like to imagine the focus here described as a kind of aesthetic version of Yan Hui's 'sitting and forgetting'. The poet's words and the courtyard's walls were my trusted accomplices in encouraging me to eliminate distractions in finding out what sort of place the garden should be.

The psychologist whose works I have found most inspiring is James Hillman (1926–2011), who has written illuminatingly on the vital 'reflexive turn' of Psyche, or the soul herself.[36] Hillman invites us to reimagine what we mean by 'soul' and to attend to its conscience, the specific logos of the psyche, rather as Pascal appealed to the specific logic – or 'reasons' – of the heart.[37] Both Hillman and Pascal are admirers of the 'reason' whose overbearing authority they nonetheless seek to set aside. For Hillman, attentiveness to soul entails a devotion to such life-changing concerns and experiences as remembrance, affliction, loss, beauty, gratitude, depth and depression. Soul is as intuitively disposed to befriend bewilderment as reason is to cast it aside.

As for the courtyard's divine sponsor, Hermes, who gives us the word hermetic and is known to be the 'guide of souls', his sanction of withdrawal in a variety of contexts proved to be an indispensable encouragement for the making of the garden.[38] The day when Hermes let it be known the courtyard and its reflex were his was a decisive turning point in the evolution of the garden.

Hermes is an unusual god, who befriends the virtue of bewilderment, which he often provokes, especially when he senses our understanding of the world has become too predictable and literal. He deprecates not matter of fact as such but our univocal and unimaginative allegiance to it. For Hermes, the visual and material world is not sufficient in itself.[39] As Hermes *Polytropos*, he is the god of 'multiple designs and aspects';[40] he reveals the double nature of things. He invites us to value both here and elsewhere, by both keeping us in the world and leading us out of it at the same time.[41] His presence is uncanny.[42] He arrives and he vanishes like the brushing of a wing. He startles us out of sorts and into bewilderment. As Hermes *Kataibates*, he is the 'one who goes down';[43] he escorts the soul to Hades after death. Understanding how bewilderment wounds us, and touching us where we are most afflicted,[44] he communicates deftly with the side of our natures that we are most apt to run away from, or, as Heraclitus puts it, 'you will not find out the limits of the soul by going, even if you travel over every way, so deep is its report'.[45]

At the centre of the Hermes courtyard is the Labyrinth of the Broken Heart, laid out as a pattern of stones on the ground in the shape of a question mark and a (musical) ear, with a dark-mouthed pot placed between them, a kind of oracle, which brings to life Hermes' nature as 'the magic darkness in bright sunlight' (Figure 8.2).[46]

FIGURE 8.2 The Labyrinth of the Broken Heart in the Hermes Courtyard, Plaz Metaxu. Source: Author

The enigmatic coexistence of light and dark is also referenced in a Georg Trakl inscription which says: '(O wie ernst ist der Antlitz der teueren Toten) / Doch die Seele erfreut gerechtes Anschaun' '(O how intent is the gaze of those who died and were dear to us) / Yet the soul enjoys this just way of looking'[47] – that is, the just apprehension of beauty will face without bias towards both life and death. Both oracle and quotation imply a rejection of binary oppositions between light and dark, being and nonbeing, life and death.

I am sometimes asked 'why is the heart broken?' To which I might reply, with Hermes, Pascal, Lorca, Meng Jiao or Sa'di:[48] 'Why would it not be?' Why would we want to repress our bewilderment given the reality we find ourselves in? One of the themes the courtyard enacts, appealing to the prayer of music in a space of the utmost forbidding constraints, is the 'Song at Closed Doors' (in poetry, since time immemorial, the topos of the *paraclausithyron*, in which the rejected lover is barred entry to the house of the beloved).[49] Claude's painting 'The Enchanted Castle' or 'Landscape with Psyche outside the Palace of Cupid' translates the theme into a mythic landscape; and at Plaz Metaxu the theme of disappointed love is reiterated outside the courtyard in Psyche's *outremer* longing for Eros across the lake, where the torment of an unhappy obsession is somewhat attenuated by the new spatial context of *l'amour de loin*, or love from afar (Figure 8.3; Garden Plan G, 14).[50]

FIGURE 8.3 The *outremer* longing of Psyche, represented as a stainless steel 'alarm sail' (*Notsegel*) (Plan G), for the domain of Eros on Orexis (Plan 14) across the lake, Plaz Metaxu. Source: Author

We gardeners think of walls as protective environments, which they are, but they are agents of confinement and defiance, too. They can lower the curtain of the night[51] and taunt us with riddles, like the sphinx.[52] Orpheus, like Hermes, is familiar with the Underworld, where he goes to recover his dead wife, Eurydice. It has been said that when he 'descends toward Eurydice art is the power that opens the night'.[53] What a wonderful image of the flowering of bewilderment! Might this not be the privilege of gardens, too, as, when here, Hermes introduces into the very sanctum and paradise of enclosure his defamiliarizing and paradoxical take on things?

Between the dark question and the listening ear of the Labyrinth is the gap where the heart is broken, the liminal space, where the unexpected transmission can occur, the uncertain, provisional and opportune 'space between', after which the garden is named. Liminality and bewilderment go together: in rites of passage, the liminal moment is the one where an old identity is abandoned before a new one can take its place.[54] It is a moment of uncertainty and danger, yet the occasion of possible transformation. Bewilderment and transformation are also closely related. Hermes delights in creating these unstable conditions, which are also inherent in Buddhist notions of transience[55] and Daoist references to the Great Transformation (*da hua*). For Sufis like Ibn 'Arabi, Rumi and Sa'di as well, there are few greater favours than to be in a state of profound religious bewilderment, which here results less from 'being at a loss' than from being found by the presence of God.[56] We have seen how the philosopher Lee Yearley proposes that we should accept bewilderment as a virtue,[57] just as we accept the virtue of ease, and he finds this virtue is necessarily and characteristically exemplified not by philosophers, but poets (giving as his examples Dante and Du Fu).[58] I would like to think a garden is – or can be – as much the friend of bewilderment as it is of contentment. In then allowing for frequent and pervasive exchanges between disturbances and composures of the heart, it refines and deepens the conscience of equanimity.

The introspections of the courtyard were foundational to the making of the garden, but could also become overbearing and oppressive.[59] The extramural development of the garden flowed above all from a complementary gratitude for space, the element in which we move and breathe freely and have our being, which is so basic to what it means to be alive that we almost universally take it for granted. Each part of the garden remembers and pays homage to space in a different way. I already mentioned how a 'reflexive' space can encourage us to make our own 'inward turn', while a 'votive' one helps us to adapt with humility towards our wider surroundings. Another example would be the 'kenotic' space of Hades (from the Greek word *kenos*, meaning empty)[60] (Garden Plan 18). In this case, the empty circular space surrounded by a yew hedge invites us, through a process of contagion, to become as empty as it is itself. In all these instances, space is valued as being our active partner rather than being disrespected as the mere blank setting for our activities, for objects and features, and even for plants.

In the main garden around the lake, at the heart of the valley, space is valued for its openness and freedom (the main lawns are dedicated to Kairos, who was the god of the right moment, bringing the gift of opportunity) (Figure 8.4; Garden Plan 13).[61] And also for its poise, as variously configured in the areas known as Epidauros, Orexis and Ananke (Garden Plan 10, 14, 15), where the garden's spatial duty of care is respectively broached, or cradled, or reconciled with the inevitable sense of an ending.[62] The main garden also serves to awaken the conscience of the valley *Gestalt*, by attending closely to the fateful spatial themes, inherent in the valley's constitution, of depth and height, and of axiality and bilaterality. Foregrounding spatial awareness in this way is, for me, a garden's central commission and gives a vital clue to how pervasively it cooperates with both nature and culture.[63]

Before, though, considering the valley at large, I would like to introduce you to the little grove of Corenzuela in the North Wood (Figure 8.5; Garden Plan 31). Corenzuela provides us with a covert vantage onto the lake. It is both a refuge and a lookout, sharing therefore in the happy spatial ambiguities, both protective and open, which Gaston Bachelard associated with the 'poetics of the corner'. Bachelard identifies the corner as 'the chamber of being'.[64] And certainly, in Corenzuela, the garden comes to provisional rest in this glade of vigilant seclusion. The 'interassimulation' of gratitude and bewilderment is also brought to fulfilment here with the help of St John of the Cross, of Chiron, the wise centaur's, cave, and of Orpheus' deft and nimble passage between familiar and numinous worlds.

FIGURE 8.4 The Kairos lawns at the centre of the garden (Plan 11). The 'axial' stream now runs beside the lake, Plaz Metaxu. Source: Author

FIGURE 8.5 The view towards the lake from the Corenzuela glade, Plaz Metaxu (Plan 31). Source: Author

Corenzuela is named after a priory in the Andalusian countryside, where Juan (de la Cruz) spent some of the happiest months of his life after suffering extreme abuse and degradation at the hands of his own Carmelite order, who had imprisoned him in Toledo.[65] We are told Juan loved whenever possible to secrete himself in tiny belfry rooms or secluded grottoes with views out over wide expanses of countryside.[66] I am moved by this predilection of the author of *The Dark Night of the Soul*[67] for wide open spaces glimpsed from a vantage of privileged confinement, exemplifying the soul's need for poise between profound inwardness and limitless expanse. This, then, is a first association of the garden's Corenzuela look-out.

A second is the memory of Chiron's cave. Chiron, the noble centaur, was tutor to the heroes, and renowned alike for his wisdom and his powers of healing.[68] But one day, in a terrible accident, Chiron was shot by his friend Herakles with a poisoned arrow, which split his luminous nature, combining the virtues of man, god and horse, apart. Neither able to heal himself, nor to die (for Chiron was immortal), Chiron retired in despair to his cave, formerly a holy place and now an accursed one, living out the perpetual plight of the human condition, for, as Hölderlin says in his poem for Chiron, 'when one is / Two in shape . . . nobody / At all knows what would be for the best'.[69]

The Orphic association of the Corenzuela glade is brought out by a Rilke inscription, written on two stones that frame the view from the lookout onto the lake (Plate 6). The inscriptions read: 'Say to the constant earth: I'm flowing' and 'To the running water speak: I am'.[70] These are the final two lines of the last of Rilke's 'Sonnets to Orpheus'. I sometimes describe the garden's way of being in the world as characterized by an 'Orphic Spatiality'. When Rilke asks if Orpheus belongs here ('Ist er ein Hiesiger?') his reply is 'no', not exclusively so, for 'his wider nature extends through both realms' (of the living and the dead, of the earth and the Underworld).[71] The garden, too, as 'the place that is between', perpetually migrates between alternative realities, and the Corenzuela glade is the privileged recess where, at our leisure, we can contemplate or participate in these migrations. It is where we can allow many dialogues to come into play: between gratitude and bewilderment, the prospect and the refuge, light and dark, the sensuous and the visionary, the inward and the expansive, the secret and the manifest, the remedy and the wound, the creaturely and the illuminated, the finite and the immortal, the open and the closed. Together, these affective and spatial dispositions, which can scarcely be separated from each other, help identify the vocation of the Corenzuela glade in the garden as the belvedere that compensates the duplex cast of things.

The Corenzuela glade gives onto the valley from an undercover refuge looking towards the lake and sunlit lawns. But what of the valley in and of itself? Over time the valley came to have a threefold meaning for me, a topographic, an imaginal and an apophatic one, each of which I will now try briefly to describe. Once again, for me, the valley as a whole was a cradle, a wound, a source of wonder and affliction, a place of belonging and not belonging. It gave shape to, and welcomed, both gratitude and bewilderment.

The valley provides the garden with its physical setting, occupying the space between two hillsides. The valley is the garden's cradle and vessel (Plate 7; Garden Plan 14). The valley topography has distinctive inclinations, including, importantly, the tendency to converge on depth while remaining open to height. The valley also, being bilateral, provides scope for reciprocal exchanges between its two sides; yet this mutuality is held in counterpoint, or even contradicted, by the linear axis the stream cuts through the centre of the valley. The valley is thus a meeting ground of sometimes complementary and sometimes opposed spatial dimensions and directions, which the gardener who pays attention to such things will want to learn from and explore. The valley also gives us the possibility of gardening at different elevations. All these correlative or divergent perspectives are somehow held together by the overarching and unifying feature of the comprehensive, symmetrical valley *Gestalt*. Such was the framing landform of the garden I was lucky enough to inherit.

In a letter written to his brother George in America in April 1819, the poet Keats, when only 23, and having recently nursed his younger brother Tom through a fatal illness, dissociated himself from those who tritely designated the world as a 'vale of tears', announcing instead his alternative belief in the world as 'The Vale of Soul-making':[72]

> I say 'Soul-making' Soul as distinguished from Intelligence . . . Do you not see how necessary a World of Pains and troubles is to school an Intelligence and make it a Soul? A place where the heart must feel and suffer in a thousand diverse ways!

This notion of the Vale of Soul-making was taken up by James Hillman in his concerted attempt to undo what he saw as centuries of European oppression of the soul, dating back, long before Cartesian dualism, to early Councils of the Church (Nicaea in 787 and Constantinople in 869–70) which progressively sought to eliminate any mediating term or influence between spirit (logos) and body (and thus any room for nourishing, ambivalent connections between rational/spiritual and physical certainties).[73] Keats's phrase, and Hillman's rehabilitation of it, became very important for the garden in imagining the valley not just as a topographical but also as a psychological vessel.

The psychological vessel has a background of woundedness (for, as is said, 'wounds are the parents of the soul') and needs piecing together with infinite care and patience.[74] It is composed of an imaginal reality that connects the spiritual and physical worlds, and forms, in Ibn 'Arabi's word, a *barzakh*, or porous boundary between them.[75] It is in this realm that the concerns of the soul, which are elsewhere arrogated by spiritual and intellectual presumption or shelved by habitual inattentiveness, can be allowed to manifest and flourish. A garden itself is like a *barzakh* between the human and natural worlds, and this alone makes it precious. When it becomes an imaginal vessel as well, when the valley becomes a Vale of Soul-making, then the concerns of the soul, as expressed for instance in poetry and myth, become a companion reflex function of, as it were they form the silver lining to, the gardener's participation in natural and spatial dwelling.

And so the garden at Plaz Metaxu is composed of spaces instinct with poetry and myth.

This vision of the garden as a Vale of Soul-making nourished me. I needed the garden, without losing its closeness to the breathings of the earth, to become at the same time what is called in Germany a *Seelenlandschaft*, a landscape of the soul. This allowed me as a gardener to fulfil both my bewilderment and my trust. At times I was almost overwhelmed with gratitude for the topopoetic correspondences which emerged. Perhaps, for others, though, they are superfluous to the requirements of a garden. In that case, I did not want to offend the garden's visitors, so I scored them unobtrusively on the ground. But for me the spaces and the gods they belong to (or is it the other way round?) co-arose in a way that never felt like a construct I was wilfully imposing on the valley. On the contrary, I felt I was receiving the valley's permission (listening to the 'request of the land') in everything I did.[76]

Perhaps it would be appropriate to think of the Vale of Soul-making as a Western inspiration of the garden. I was working, after all, within the tradition of Keats and Hillman, for whom the Greek gods remained of vital concern in the making of our modern psyches and behaviours. This imaginal layer of the garden appeared to me contemporary, complex and immanental: the gods got under the garden's skin as a matrix of furtive inspirations, reflecting our provisional, heterodox and fallible intuitions about our place in the world. The garden's identity as a Vale of Soul-making, as we might expect from Keats's 'thousand diverse ways', is a richly counterpoint and manifold one.

Nonetheless, when we are at the centre of the garden, the main impression we have is one of openness and simplicity, a placid scene of lawn and lake, as unguarded as it is serene. The transparent composure of the landscape here offsets and balances, and even unravels, the intricate and complex relationships at play in other areas of the garden. Often we will want to take this serenity at face value – I frequently do myself – and that is altogether appropriate, producing its own sense of calm and gratitude. However, in turning to the third level of meaning the valley carries for the garden, this serenity is not a constant given, for it can mask a more mysterious and initially forbidding aspect: this is when the physical and imaginal reality of the garden gives way to the numinous apprehension of the valley as an apophatic, or decreative, vessel that works to unmake what has been made and to cause to vanish what has been ostensible. This gesture has many resonances, but perhaps chief among them for me were Eastern ones, typified by the image of the valley as intrinsically given to unmaking or undoing itself (a kind of imaginal parallel to the self-abrogative function of language in apophatic discourse) as if in an implicitly perpetual – and inexhaustible – performance ('the valley spirit that doesn't die'[77]) of the well-known saying in the Heart Sutra that 'form is emptiness'. Of course, the Heart Sutra is not one-sided, and immediately adds that 'emptiness is form'[78]. Abrogation and emergence go together. In that case my emphasis on the valley's decreative gesture here is certainly a distortion of the overall picture. Given Western culture's overwhelming bias towards positivistic descriptions and appropriations of phenomena, I nonetheless think the distortion is justified (the material valley will always reappear soon enough – have no fear!). For me, on reflection, or simply through intent noticing, the simplicity and openness of the valley's presence can be experienced as housing an empty space which, when diligently attended to, turns into an intimate invitation to us to discover our own constitutional emptiness. If this invitation is unsettling at first, with practice it becomes absolving and befriending.[79]

This invitation is further articulated by a curved caesura cut into the lawn at the heart of the garden, near the shallow end of the lake, which is inscribed with a quotation from Terence, denoting acceptance of the human world, and another one of Sufi inspiration, imagining the world we live in as one arising from the sigh of a compassionate god (Figure 8.6. Garden Plan E).[80] The idea is that within the between space of either the valley or the caesura we are invited to place ourselves with the utmost humility at the disposal of

FIGURE 8.6 The caesura (Plan E) set in the lawn at the east end of the lake.

both the mortal and the immortal, or, if we prefer, the preternatural, horizons of our lives. To enter this empty, liminal space requires us to shed the props of our habitual identity. I was very moved one day when Yue was giving us a seminar on the Garden of Artless Administration, in Suzhou, when she showed us the eighth of Wen Zhengming's famous views of the garden, where a man is sitting by the edge of a pond at what is entitled 'Purifying Will Place'.[81] The accompanying poem draws out the connection between the name and the image in such a way as to confirm my own intuitions about the ways in which a spatial configuration can interact with and influence a psychological one – in this case, how an empty between space, whether cradled at the heart of the valley or within the caesura, can appeal for a liminal interruption of discursive thought processes, so provoking an unconditional readiness for transmission. Another of Wen's views, 'The Thoughts of Afar Terrace',[82] dramatically captures the kind of precipitate suspension the between space can invite: is it far-fetched to suppose we are here presented with a spatial representation of the kind of experience Yan Hui, as we saw, refers to as 'letting my mind and body fall away' so as to 'make myself identical with the Great Thoroughfare'?

In Sufi terms, this abrogative function of the vessel of the valley could be likened to cultivating the state of *fana* ('passing away' or 'annihilation') in which the ideal is to learn how to die before one dies; whereas, in Western terminology and practice, it has affinities with the *via negativa* (or negative way of knowledge of, or union with, ultimate reality).[83] As with everything else at Plaz Metaxu, extreme positions like these are presented not as arbitrarily insistent rules for how to read the garden, but as tentative options of interpretation arising spontaneously from the intrinsic nature of a 'place that is between'. Interestingly, on the aesthetic level, Keats also favoured a 'negative capability' which he describes as 'being in uncertainties, Mysteries, doubts, without any irritable reaching after fact & reason'.[84] All these apophatic, decreative or negative perspectives are, I suggest, like the deepest affinities of the soul, unusually *yin* or bewilderment friendly, in their willingness to work with mystery, or to let mystery resonate through from the margins to the centre of life. In a garden, or my garden anyway, this working with mystery is expressed as the willingness to dwell liminally, that is, the willingness to place oneself through one's surroundings at the disposal of the liminal moment, which entails allowing bewilderment to flourish, when it needs to, alongside contentment, gratitude and trust.

Notes

1 Han Yu (768–824), *Preface sent to Meng Jiao* (*Song Meng Dongye Xu*) quoted by Duncan Campbell in 'The Cultivation of Exile: Qi Biaojia and his Allegory Mountain' in *Creativity in Exile*, ed. Michael Hanne (Amsterdam: Editions Rodopi B.V. 2004; Brill, 2004/2016), 255.

2 See David E. Cooper, 'Garden practices, "quiet" virtues and understanding', Chapter 6 in this volume.

3 See Lee H. Yearley, 'The Ethics of Bewilderment', *The Journal of Religious Ethics*, 38, no. 3 (September, 2010): 436–60.

4 For the relationship between a primal 'porosity' and a 'passivity' of being, see William Desmond, *The William Desmond Reader*, ed. Christopher Ben Simpson (Albany: State University of New York Press, 2012), 145, 203.

5 Lionel Trilling, 'Wordsworth and the Iron Time', *The Kenyon Review*, 12, no. 3 (Summer, 1950): 477–97.

6 Perhaps the most frequently heard comment by visitors to my garden regards its tranquillity, which is clearly perceived to be a desirable and characteristic virtue not just of my garden, but of gardens generally. The tranquillity of a garden is indeed a virtue, but it is not a constant. It clearly accommodates agreeable sounds such as birdsong or rustling leaves in a breeze, but not more agitated (and agitating) weather conditions. A garden is subject to many external disturbances, some of them violent, and to many internal ones, too, namely the numerous worries and disappointments that inevitably accompany the exercise of care, which applies in an environmental context as much as it does in a personal one. I myself want to resist the inclination to consider the privileged tranquillity a garden can offer us in isolation from the more troubling influences a garden is heir to. Because a garden may provide us with a retreat from one set of circumstances, this does not mean there are not other circumstances (sometimes more far-reaching ones) it requires us to face with

trepidation on a daily basis. In my experience, equanimity and trepidation are interdependent in the garden. The equanimity of which I am capable there is not a stable immunity to apprehension, nor a delusive defence against it, but the provisional realization of a successful reabsorption of trepidation into a reconfigured moment of calm and serenity, through a never-ceasing process of openness and adaptation. Equanimity is never a given, but always a product of dynamic attentiveness. For a useful discussion of the distinction between mindfulness and equanimity in a Therevada Buddhist context, see Gaëlle Desbordes et al., 'Moving beyond Mindfulness: Defining Equanimity as an Outcome Measure in Meditation and Contemplative Research', in *Mindfulness (NY)*, 6, no. 2 (April 2015): 356–72. Mindfulness, which is already here dissociated from ordinary or random attentiveness as 'the heightened awareness of present moment sensory, affective and cognitive experiences', is itself then seen as no more than a rehearsal for the 'complex construct' of equanimity, 'a mental state or trait . . . [of] even-mindedness in the face of every sort of experience . . . that is not easily achieved and typically requires some form of practice'.

7 In modern usage, the concept of liminality became current with Arnold van Gennep's discussion of tribal rites of passage in 1909 in *The Rites of Passage*, trans. Monika B. Vizedom and Gabrielle L. Caffee (Chicago: University of Chicago Press, 1961). The concept has since been more widely applied by Victor Turner and others. For two recent approaches, see Bjørn Thomassen, *Liminality and the Modern: Living through the In-Between* (London and New York: Routledge, 2014) and Mihai I. Spariosu, *Modernism and Exile: Liminality and the Utopian Imagination* (London: Palgrave MacMillan, 2014).

8 Giacomo Leopardi (1798–1837), *Zibaldone: The Notebooks of Leopardi*, ed. Michael Caesar and Franco D'Intino (London: Penguin Books, 2013), 1823 (paras 4175–7). This passage is dated Bologna, 22 April 1826.

9 Both in Europe and China the goal of equanimity has long been associated by philosophers with the theme of retirement to a garden. In Europe, the connection between gardens and contemplative virtue, or enlightened *otium* (leisure), goes back to Plato (*c.* 427–347 BCE) and Epicurus (341–270 BCE), and Roman authors such as Cicero (106–43 BCE) and Pliny the Younger (61–113 CE), while in China the private or temple garden was always associated with a practice of contemplative 'self-cultivation' deeply imbued with Confucian, Daoist and Buddhist ideas. For the Roman 'otium villa', see James S. Ackerman, *The Villa: Form and Ideology of Country Houses* (Princeton: Princeton University Press, 1993), ch. 2 'The Ancient Roman Villa', 35–61. For the origins of the Chinese tradition of self-cultivation, see Paul Fischer, *Self-Cultivation in Early China* (Albany: State University of New York Press, 2023). For the tradition of self-cultivation in Chinese gardens in Suzhou, see Donia Zhang, 'Classical Chinese Gardens: Landscapes for Self-Cultivation', in *Contemporary Urban Affairs*, 2, no. 1 (2018): 33–44. Equanimity in the Zen garden is premised on the experience of awe and withdrawal from habitual, discursive thought processes.

10 In Chinese, *wan wu*. This term, first found in *Daodejing*, ch. 42, became a standard way of referring to everything that exists in classical Chinese thought.

Until recently, when Western ideas started to be introduced into Chinese philosophy, there was no Chinese word for 'nature' as such. Instead, terms such as *tian* (heaven), *tian di* (heaven and earth) and *zi ran* (spontaneous occurrence) were used. The latter term has now evolved to be the closest equivalent of 'nature' as understood in the West.

11 Yan Hui (*c.* 521–481 BCE) was the favourite disciple of Confucius. As he appears in the *Zhuangzi*, probably written in the fourth century BCE, he takes on the character of a Daoist sage, who surpasses Confucius in his enlightened apprehension of the Way. See *Chuang-Tzu: The Inner Chapters*, trans. A. C. Graham (Indianapolis/Cambridge: Hackett Publishing Company, Inc., 1981).

12 Tao Yuanming (365–427), also known as 'hermit Tao' (Tao Qian or T'ao Ch'ien), was a celebrated Chinese poet, who founded the tradition of 'fields and gardens' poetry, and lived on a modest farm in the vicinity of Mount Lu, south of the Yangtze river near the city of Jiujiang. After several failed attempts, he finally succeeded in leaving government service and lived with his family in near total seclusion.

13 William Wordsworth (1770–1850), major English Romantic poet, who retired with his sister Dorothy to Grasmere, in the Lake District, in 1799.

14 For a discussion of the dialogue between Yan Hui and Confucius in ch. 6 of the *Zhuangzi*, see Edward Slingerland, *Effortless Action: Wu-wei as Conceptual Metaphor and Spiritual Ideal in Early China* (Oxford: Oxford University Press, 2003), 185. See also A. C. Graham, *Chuang-Tzu*, 92.

15 See Harold D. Roth, 'Daoist Apophatic Meditation: Selections from the Classical Daoist Textual Corpus', in *Contemplative Literature: A Comparative Sourcebook on Meditation and Contemplative Prayer*, ed. Louis Komjathy (Albany: State University of New York, 2015), 89–143.

16 Geoffrey H. Hartman, 'A Poet's Progress: Wordsworth and the *Via Naturaliter Negativa*', in *Modern Philology*, 59, no. 3 (February 1962): 214–24.

17 Among modern commentators, A. R. Davis and Yim-tze Kwong are united in speaking of Tao's 'essentially melancholic nature', while Lin Wen-yüeh also came to detect beneath the tranquil and peaceful surface of the poetry 'an abiding sense of loneliness' (quoted in Yim-tze Kwong, 36). See A. R. Davis, *T'ao Yüan-ming: His Works and their Meaning*, 2 vols (Cambridge: Cambridge University Press, 1983); Yim-tze Kwong, 'Naturalness and Authenticity: The Poetry of Tao Qian', *Chinese Literature: Essays, Articles, Reviews*, 11 (December 1989): 35–77.

18 'Ode: Intimations of Immortality from Recollections of Early Childhood', 144. In the Isabella Fenwick note to this poem, Wordsworth recalls that 'I was often unable to think of external things as having external existence and I communed with all that I saw as something not apart from, but inherent in, my own immaterial nature. Many times while going to school have I grasped at a wall or tree to recall myself from this abyss of idealism to the reality. At that time I was afraid of such processes. In later periods of life I have deplored . . . a subjugation of an opposite character and have rejoiced over the 'Obstinate questionings / Of sense and outward things, / Fallings from us, vanishings . . .' In *The Prelude*, bk 6, lines 53ff, Wordsworth also refers to the 'glory' of the 'Imagination' as occasioning 'such strength / Of usurpation . . . when the light

of sense / Goes out in flashes that have shown to us /The invisible world'. Quotations from *The Prelude* are from the 1805 version in William Wordsworth, *The Prelude: A Parallel Text*, ed. J. C. Maxwell (Harmondsworth, Middlesex: Penguin Books Ltd, 1971).

19 Wordsworth, *The Prelude* (1805), bk 1, line 420.

20 A 'negative' or apophatic approach to communicating or undergoing the experience and reality of the nature of God (or ultimate truth) is also well known in Christian and other schools of mysticism, for instance as advocated by the *Mystical Theology* of Pseudo-Dionysius the Areopagite (fifth–sixth century CE), the anonymous fourteenth-century author of *The Cloud of Unknowing*, or by St John of the Cross in his *Dark Night of the Soul* (1584–6). The Sufi concept of *fana* (annihilation, or passing away, of the self) as a condition of experiencing the divine is a related one and bears comparison with the less ostensibly ecstatic Buddhist notions of *nirvana* (becoming extinguished) and *śunyātā* (universal emptiness of self-nature), all of which deconstruct positivistic assumptions about identity and adopt a self-abrogative approach towards both the efficacy and the performance of language. More important than the critique of language per se is the apophatic training of the focused attentiveness within contemplative practice.

21 For 'decreation' see Simone Weil: 'Décréation: faire passer du créé dans l'incréé. – Destruction: faire passer du créé dans le néant. Ersatz coupable de la décréation' (Decreation: to make something pass into the uncreated. Destruction: to make something created pass into nothingness. A blameworthy substitute for decreation) in Simone Weil, *Gravity and Grace*, trans. Emma Craufurd (London: Routledge and Kegan Paul, 1972), 28.

22 'Nature loves to hide'. Heraclitus, Fragment X, in *The Art and Thought of Heraclitus: An Edition of the Fragments with translation and commentary*, ed. and trans. Charles H. Kahn (Cambridge: Cambridge University Press, 2008), 33.

23 Buddhism and Daoism are far more sympathetic to advocating the reciprocal virtue of the (cosmological) gesture of withdrawal (in relation to the gesture of emergence of life) than are most Western philosophies. I personally think this more comprehensive intuition, which envisions ontology as a reversible rhythm between coming into, and receding from, being, provides a major vindication for the perspective of a life of contemplative withdrawal, whether in the garden or anywhere else.

24 Liu Yuan in Suzhou is actually called 'The Lingering Garden' or 'The Garden that Lingers between Heaven and Earth'. A *karesansui* (dry landscape garden) like that at Ryoan-ji in Kyoto both distils the energies and the forms and voids of the cosmos into a momentous, miniature 'islands-and-ocean' representation and inclines us towards the meditative state of mind for which such distillations are, of necessity, a second nature. A rare kind of reciprocal equivalence is reached between the reality of cosmic appearance and behaviour on the one hand and the transforming influence of contemplative withdrawal and projection on the other.

25 The Park and Château of Courances, some 60 km south of Paris, date from the sixteenth to the eighteenth centuries. Courances is perhaps the most beautiful of all French water gardens, distilled and intimate in character, and without the

usual rhetorical embellishments of extravagant fountains, prominent statuary and formal gravel or sandy walkways, relying instead entirely on the simplest elements of garden design: sheets of water, expansive lawns and hedge-lined areas of woodland.

26 Studley Royal was laid out by John Aislabie between *c.* 1715 and 1742. The Georgian water garden, an exquisitely formal, yet tactful, intervention in the rugged valley of the River Skell, is among the most beautiful in Europe. As at Courances, but in a much more remote and hilly setting, the garden employs the simplest elements and clearest geometries to complement its natural surroundings in a quietly sublime and revelatory way.

27 A *ricetto* is a fortified sanctuary found in Italian villages. It is also the name by which Michelangelo's notoriously perverse and claustrophobic vestibule to the Laurentian Library in Florence is familiarly known. Many artists, including Rothko, have been haunted by this radically introspective and brooding architectural space, whose poetic presence has been variously described as 'oppressive, steep [and] overpowering' and as expressive of 'insoluble conflict'. Yet the *Ricetto* also functions as a necessary antechamber to the luminous calm of the Bilblioteca Laurenziana itself. For excellent discussions of the *Ricetto*, see Ludwig H. Heydenreich and Wolfgang Lotz, *Architecture in Italy 1400–1600* (Harmondsworth: Penguin Books, 1974), 248; Rudolf Wittkower, 'Michelangelo's Biblioteca Laurenziana', *Art Bulletin*, 16, no. 2 (1934): 123–218; James S. Ackerman, *The Architecture of Michelangelo*, vol. I, ch. IV, 'The Library of San Lorenzo' (London: A. Zwemmer Ltd, 1966), 38–44 esp. In this way, too, it serves as a model in the garden where a similar heightened gratitude, in this case for space at large, is prepared for by the fraught confinement generated within the 'Distress Retort'.

28 Dante famously begins the first canto of the *Inferno* with the words: 'Nel mezzo del cammin di nostra vita / mi ritrovai per una selva oscura, / ché la diritta via era smarrita' (Midway in the journey of our life / I came to myself in a dark wood, / for the straight way was lost). Dante Alighieri, *Inferno* I, 1–3, trans. Robert Hollander and Jean Hollander (New York: Anchor Books, Random House, Inc., 2002).

29 Evelyn Underhill, *Mysticism: A Study in the Nature and Development of Man's Spiritual Consciousness* (London: Methuen & Co. Ltd, 1911), 4.

30 Just as 'contemplation is the mystic's medium', so 'introversion is the characteristic mystic art'; Underhill, 298. This characteristic is by no means restricted to Christian mysticism, Underhill's main focus of study. From the Delphic maxim 'Know thyself' to Dōgen Zenji's teaching that 'To study the Buddha Way is to study the self', the advocacy of a dedicated introspection (in Daoism *neiye*, or 'inner training') is a constant at the heart of religious contemplative practices worldwide. Not to be confused with 'navel-gazing', or what Louis Lavelle, in his book on Narcissus, calls the 'bad' or 'wrong' type of introspection, such 'good' or 'wise' introspection leads to a self-forgetfulness (Dōgen: 'To study ourselves is to forget ourselves'), which provokes in us a profound hospitality towards the 'myriad things' (Dōgen) or 'the representatives of the normative order' (Zhuangzi), or, as Lavelle puts it, ' a world opens out which is within us but into which every being can be invited'. Such an experience of 'elemental intimacy' with the world, far from sealing off

true contemplatives from their surroundings, invokes an unusual sympathy with them. So in Buddhism, for instance, the goal of enlightenment also gives rise to an active compassion for all creatures and things. Yet an inner training, almost always dependent on a period of intense withdrawal, which cultivates an 'air-tight' seal between the inner and the outer worlds, is also a pre-requirement for the equanimity and freedom from 'entanglements' of the self-possessed and enlightened sage and/or dedicated mystic. So it was said of Liezi that, for a period of three years, 'In the face of entanglements he remained sealed (*feng*), and in this oneness he ended his life', while for Zhuangzi, too, it is precisely the emptying of the container of the Self of extraneous elements, and the keeping of it sealed against the outside, which permits it 'to serve . . . as a location where the normative order or its representatives can come to dwell'. See Dōgen Zenji, *Genjo Koan*, section 8, in *Dōgen's Genjo Koan: Three Commentaries*, trans. and comm. by Nishiari Bokusan et. al. (Berkeley, CA: Counterpoint Press, 2011), 23; Louis Lavelle, *The Dilemma of Narcissus*, trans. William Gairdner (Burdett, New York: Larson Publications, 1993), 43–4. For Zhuangzi, see Slingerland, *Effortless Action*, ch. 6, 'The Tenuous Self: Wu-wei in the *Zhuangzi*', 175–215. For the above quotations about Liezi and Zhuangzi, see Slingerland, 190.

31 In using the word 'impossible' here, I might signal a few resonances the term has for me. Thus I admire how Zen masters are unafraid to represent themselves as friends of the impossible, as when rōshi Shunryu Suzuki writes that '"although Buddhism is unattainable, we vow to attain it." If it is unattainable, how should we attain it? But we should! That is Buddhism. To think, "Because it is possible we will do it" is not Buddhism. Even though it is impossible, we have to do it because our true nature wants us to.' Shunryu Suzuki, *Zen Mind, Beginner's Mind: Informal talks on Zen meditation and practice* (New York: Weatherhill, Inc. 1970), 45. On a more poetic level, I might recall Rilke's use of the word 'Unmögliche' (translated by Spender/ Leishman as 'impossibles') in the first Duino Elegy. The passage reads: 'Voices, voices. Hearken, my heart, as only / saints once hearkened: so, that, the giant call / lifted them off the ground; they, though, impossibles, / went on kneeling and paid no heed, / such was their hearkening.' Rainer Maria Rilke, *Duino Elegies*, The First Elegy (1922), trans. J. B. Leishman and Stephen Spender (London: Chatto & Windus Ltd, The Hogarth Press, 1939), 53–7. I am also moved and impressed by how writers like Kafka, thinkers like Pascal and philosophers like Kierkegaard do not flinch from drawing existential consequences from the encounter with paradox or 'tragic impossibility', allowing their commitments to 'futile unhappiness', to 'the reasons of the heart' or to 'the teleological suspension of the ethical' to expose the questionable truth-status of consensus beliefs in rationality or morality as insufficiently permeable to the reality of the unaccountable. Another resonance for me would be the following passage from Thomas Browne's *Religio Medici*: 'As for those wingy mysteries in Divinity, and ayery subtilties in Religion, which have unhindg'd the braines of better heads, they never stretched the *Pia Mater* of mine; me thinkes there be not impossibilities enough in Religion for an active faith . . . I love to lose my selfe in a mystery to pursue my reason to an *oh altitudo* . . . I can answer all the Objections of Satan and my rebellious reason with that odd resolution I learned of *Tertullian, certum est quia impossibile est*

(It is certain because it is impossible). I desire to exercise my faith in the difficultest points.' Sir Thomas Browne, *Religio Medici* I.9 (1643) in *Religio Medici and Other Works*, ed. L. C. Martin (Oxford: The Clarendon Press, 1964), 9.

32 The term first appeared in Ortega's essay *Ensimimasmiento y Alteración*, later incorporated as the first chapter of his book *El Hombre y La Gente* (Madrid, 1957), translated by Willard R. Trask as *Man and People* (New York and London: W. W. Norton & Company, Inc., 1963). For Ortega, the act of 'withdrawal' was the individual's necessary first step of independence against 'alienation' and inauthenticity.

33 T'ao Ch'ien, *Home Again Among Gardens and Fields* I, 1–2, in *Selected Poems of T'ao Ch'ien*, trans. David Hinton (Port Townsend: Copper Canyon Press, 1993). For the different versions of Tao's name, see note 12.

34 Emily Dickinson poem # 1072 in *The Poems of Emily Dickinson*, ed. R. W. Franklin (Cambridge, MA: Harvard University Press, paperback edition, 2005).

35 Ibid., poem # 409.

36 'By *soul* I mean, first of all, a perspective rather than a substance . . . This perspective is reflective.' James Hillman, *Re-Visioning Psychology* (New York: Harper & Row, 1992 [1975]), xvi.

37 Blaise Pascal, *Pensées* (1669), introduction by T. S. Eliot (New York: E. P. Dutton & Co., 1958), section 4, para 277, 'Le coeur a ses raisons que la raison ne connaît point' (The heart has its reasons of which reason knows nothing).

38 Karl Kerényi calls his monograph on Hermes *Hermes, Guide of Souls: The Mythologem of the Masculine Source of Life*, thus emphasizing Hermes' role as psychopomp, trans. Murray Stein (Dallas: Spring Publications, 1992 [1976]).

39 The Hermetic way of seeing is unimpressed by daylight and the instrumental eyes it gives us. Hermes hears, whispers, flies, imagines, touches, vanishes – all without recourse to ordinary sight.

40 'Polytropos' is an epithet Hermes shares with Odysseus. For a full list of Hermes' epithets, see William G. Doty, 'Hermes' Heteronymous Appellations', in *Facing the Gods*, ed. James Hillman (Spring Publications, 1980).

41 'This is Hermes . . . the God of betweens, keeping us to the world and guiding us out of it at the same instant.' Hillman, 'Notes on Opportunism', in *Puer Papers*, ed. James Hillman (Dallas: Spring Publications, 1989 [1979]), 156.

42 'Everywhere that Hermes appears . . . there is an influx and invasion from the underworld.' Kerényi, *Hermes, Guide of Souls*, 85.

43 The topos of descent is a recurrent theme in the garden and very much part of its character as a place that is between. Descent and depth, close associates of bewilderment, are less readily explored than ascent and height. Or, as the philosopher William Desmond puts it, 'Too often we think of the between as a space from which to go up – but what about going down into the deeper darkness?' Desmond, *The William Desmond Reader*, 100.

44 'Hermes is a friend in the desperation of our solitude.' Raphael López-Pedraza, *Hermes and His Children* (Einsiedeln: Daimon Verlag, 1989), 24.

45 Heraclitus, Fragment XXXV, in *The Art and Thought of Heraclitus*, 45.

46 Walter F. Otto, *The Homeric Gods: The Spiritual Significance of Greek Religion* (1929), trans. Moses Hadas (London: Thames and Hudson, 1979), 118.

47 The quotation is from Trakl's poem Helian I, 20–1, translation mine. The most comprehensive bilingual edition available of Trakl's work is by Jim Doss and Werner Schmitt, *Georg Trakl: The Last Gold of Expired Stars Complete Poems 1908–1914* (Sykesville, MD: Loch Raven Press, 2010).

48 Blaise Pascal (1629–62), French mathematician, thinker and Christian apologist, who explored the tragic paradoxes of life according to the conscience of the 'heart's reasons'. Federico García Lorca, Spanish lyric poet and dramatist (1898–1936). Meng Jiao (751–814), a Tang dynasty poet, in the circle of Han Yu (see note 1), who broke with tradition by exploring the dark and disconsolate face of nature. Sa'di of Shiraz (1210–91/2), a major Sufi poet, who pronounced on his return to Shiraz after years spent studying abroad, 'For years his intellect left – he thought – to learn tranquillity: / What did it learn? As more bewildered than ever it has returned. // Behold how the intellect escapes from the flood of pain of love: / At large for years across the world, and yet to the whirlpool of danger it has returned.' The full translation of this ghazal is given by Domenico Ingenito in *Beholding Beauty: Sa'di of Shiraz and the Aesthetics of Desire in Medieval Persian Poetry* (Leiden, Boston: Brill, 2021, available as a free pdf download), 4–6. The critique of the 'binding' intellect (*'aql*) by means of the more open and transformative knowledge of the heart (*qalb*) is a constant theme in Sufi discourse, as exemplified also by Ibn 'Arabi. The heart has powers of divination into reality (sometimes profoundly disquieting ones) which surpass those of the philosophical *ratio*, whether for the Sufis or for Pascal (or, of course, for most lyric poets).

49 For a discussion of this topos of the 'lament beside the door', see H. V. Canter, 'The Paraclausithyron as a Literary Theme', in *American Journal of Philology*, 41, no. 4 (1920): 355–68. The theme of the unrequited lover (*exclusus amor*) was a mainstay of Troubadour lyric poetry. The lasting importance of the interconnection between passion and the obstacles placed in its path was explored by Denis de Rougemont in his meditations on the Tristan and Isolde myth in *L'Amour et l'Occident* (1939), translated by Montgomery Belgion as *Love in the Western World* (1940) (Princeton: Princeton University Press, 1983). In contrast to the Western emphasis of de Rougemont's account, it is now thought Arabic sources, entering Provence through Spain, played a vital role in forming the Troubadour vision of love. See, for instance, María Rosa Menocal, 'Close Encounters in Medieval Provence: Spain's role in the Birth of Troubadour Poetry', in *Hispanic Review*, 49, no. 1 (Winter, 1981): 43–64 and the same author's book *The Arabic Role in Medieval Literary History* (Philadelphia: University of Pennsylvania Press, 1987).

50 Jaufré Rudel (d. *c.* 1147) was the troubadour who developed the theme of *amor de lonh* in his songs, a 'love from afar' supposedly occasioned by his longing for the beautiful, distant Countess Hodierna of Tripoli. Kaija Saariho's *L'Amour de Loin*, with a libretto by Armin Maalouf, is a contemporary opera based on Rudel's story and was spectacularly staged at the Metropolitan Opera House, New York in 2016.

51 Of the connection between confinement and tragedy, William Desmond writes: 'Tragedy reveals one of the ultimate forms of being at a loss. In a tragic situation we are faced with boundaries from which there is no escape. We are backed against a wall, as it were. We are pushed without recourse into a dead end of being. How we are, and what we do in this cul de sac, is revealing of what it is to be.' Desmond, *The William Desmond Reader*, 167.

52 In his book on 'Earth and Reveries of Will' (*La Terre et les Rêveries de la Volonté*) Gaston Bachelard supposes 'la vraie matière du sphinx c'est le rocher' (the real subject of the sphinx is rock) (Paris: José Corti, 1978), 195. Walter Burkert draws the connection with Hermes: 'It is not only the phallic sign that belongs to Hermes, but also the pile of stones . . . This interrelation is possible precisely from the signal function of both phallus and stone.' *Structure and History in Greek Mythology and Ritual* (Berkeley: University of California Press, 1979), 41.

53 Maurice Blanchot, 'Quand Orphée descend vers Eurydice l'art est la puissance par laquelle s'ouvre la nuit'. This is the opening sentence of Blanchot's essay 'Le Regard d'Orphée' in *L'Espace Littéraire*, trans. Ann Smock as 'Orpheus' Gaze', in *The Space of Literature* (Lincoln, London: University of Nebraska Press, 1982), 171.

54 See note 7 for the provenance of liminality dating back to Van Gennep.

55 For a fascinating insight into how Buddhist notions of transience can help determine the character of a garden, see Lin Zhu's essay on Qi Biaojia's garden, Mountain Yu, Chapter 1 in this volume.

56 *Hayrah* is the Sufi term for bewilderment. The being found by God (Allah) means that you live in the constant presence of the Unknowable. Rational, conceptual and all other positively freighted constructs are an impediment to surrendering to the Unknowable nature of Allah. Bewilderment, on the other hand, through disabling such constructs and opening itself to transformation, facilitates the course of such surrender. In his *Fusus al-Hikam* (The Bezels of Wisdom), Ibn 'Arabi frequently quotes the famous hadith attributed to the prophet: 'Lord, increase me in bewilderment in you'. For a recent translation of the *Fusus* see Binyamin Abrahamov, *Ibn al-'Arabiī*'s *Fusūs Al-Hikam: An Annotated Translation of 'The Bezels of Wisdom'* (Milton Park: Routledge, 2019). For a discussion of Ibn 'Arabi and bewilderment, see Michael A. Sells, *Mystical Languages of Unsaying* (Chicago: University of Chicago Press, 1994), 101–3, 108–12, and also Ian Almond, 'The Honesty of the Perplexed: Derrida and Ibn 'Arabi on "Bewilderment"', *Journal of the American Academy of Religion*, 70, no. 3 (September 2003): 515–37. For Jalaluddin Rumi (1207–73), the greatest Sufi mystic and poet in the Persian language, bewilderment comes from love, cleverness (reason) from Iblis (the devil). In the second story of Book IV of his magnum opus, the *Masnavi*, para 29, Rumi states famously, in the 1898 translation of E. H. Whinfield, 'Sell your cleverness and buy bewilderment' (available online https://archive.sacred-texts.com/isl/masnavi/msn04.htm, accessed 8 September 2024). The *Masnavi-ye-Ma'navi* (or Spiritual Couplets), to give it its full title, runs to some 25,000 verses and is often regarded as the single greatest mystical poem in world literature.

57 See note 3.

58 Du Fu (712–70) was a Tang dynasty poet, often paired with his friend Li Bai (701–62) as one of the two greatest Chinese poets.

59 The panic and sense of claustrophobic entrapment induced by confinement are more pointedly explored in the small adjacent courtyard (Plan 5), which is dedicated to Pasiphaë and Danaë, who were held prisoner by their husband and father respectively.

60 St Paul (Philippians 2.7) used the verb *kenóō* to describe the experience by which Christ emptied himself of one side of his nature. William Desmond speaks of our own need for 'a kenosis of the will . . . an emptying out of the self-insistence that clogs our porosity to what is beyond us'. 'Kenosis and generosity are twinned, for the point of being nothing is to be an everything of welcome for the other.' Desmond, *The William Desmond Reader*, 217–18.

61 Kairos is characterized in Aesop's *Fables*, written in the sixth century BCE, as 'Running swiftly, balancing on the razor's edge, bald but with a lock of hair on his forehead. He wears no clothes; if you grasp him from the front, you might be able to hold him, but once he has moved on not even Zeus himself can pull him back: this is a symbol of Kairos, the brief moment in which things are possible.' *Aesop's Fables*, no. 536, trans. Laura Gibbs (Oxford: Oxford University Press, 2008). I have substituted 'Kairos' and 'Zeus' for 'Tempus' and 'Jupiter' in the Gibbs translation. Kairos is best known to us through the private relief sculpture Lysippos (390–300 BCE) made of him. In the relief, known only through copies, Kairos carries a scales as well as a razor and has wings attached to his shoulders and ankles. See also Dietrich Boschung, *Kairos as a Figuration of Time: A Case Study* (Munich: Wilhelm Fink Verlag, 2013) available through Brill as an e-book (2019).

62 Epidauros, today known for its theatre, was an important religious site in ancient Greece, whose main sanctuary was for Asclepius, the god of healing. In the garden, the theatrical and healing components of this key space of transition are both important in generating the kind of spatial poise I associate with a garden's main duty of care towards its visitors. Orexis, from the Greek *oregō*, 'I reach', is the site in the garden of Eros' palace (represented as a stand of slender slates) and so 'connotes . . . in the inner realm . . . an [aspirational] pointing towards something'. Martha Nussbaum, *The Fragility of Goodness* (Cambridge: Cambridge University Press, 2009), 274. However, Eros' idealism is here brought into association with more fallible or wounded perspectives, as represented by the troubled *pueri* Icarus and Narcissus, not to mention the probing and subversive inclinations of Psyche herself. Orexis (see Plate 7) thus becomes a seminal site of chiasmus and peripeteia in the garden (a vital crossroads and turning-point). Height and depth, axial and transverse momentums are brought into balance as we stand on Orexis where, more than anywhere else in the garden, we become aware of the valley's reconciling and cradling action, which generates a new kind of spatial poise. Ananke was the goddess of Necessity. Her garden, below the dam of the lake, is poised between the main garden and its theatre of betweenness, where orthodox temporality is held in abeyance, and the resumption of linear temporal necessity in the outlying field of Hesperos (Plan 35).

63 Spatiality is fundamental to both gardens and nature, but is seldom valued as such. It invokes perspectives and horizons which transcend both

anthropocentrism and biocentrism. For a complementary appreciation of the value of stone, especially in the East Asian context, see François Berthier, *Reading Zen in the Rocks: The Japanese Dry Landscape Garden*, trans. with a Philosophical Essay by Graham Parkes (Chicago: University of Chicago Press, 2005).

64 Gaston Bachelard, *The Poetics of Space* (*La Poétique de l'Espace*, 1958), trans. Maria Jolas (Boston: Beacon Press, 1994), 138.

65 For the life of St John of the Cross, see Gerald Brenan, *St John of the Cross: His Life and Poetry* (Cambridge: Cambridge University Press, 1973).

66 Brenan, *St John of the Cross*, 48.

67 Juan wrote his celebrated lyric *En una Noche Oscura* (The Dark Night) in *c.* 1578–9, probably just after his escape from prison in Toledo. He wrote two extended commentaries on this poem: the *Ascent of Mount Carmel* (1579–85) and the *Dark Night of the Soul* (1584–5).

68 For Chiron, see Homer, *Iliad*, bk. 11, line 831 and Pindar, *Pythian Ode* 3.5 in *Pindar: The Complete Odes*, trans. Anthony Verity (Oxford: Oxford University Press, 2008).

69 Hölderlin, *Chiron*, in *Friedrich Hölderlin Selected Poems*, trans. David Constantine (Newcastle upon Tyne: Bloodaxe Books, 1990), 34–7.

70 Rilke, *Die Sonette an Orpheus* II. XXIX (Leipzig: Insel Verlag, 1923): 'zu der stillen Erde sag: Ich rinne. / Zu dem raschen Wasser sprich: Ich bin.' (Translation mine.)

71 Rilke, *Die Sonette an Orpheus* I. VI: 'Ist er ein Hiesiger? Nein, aus beiden / Reichen erwuchs seine weite Natur.'

72 Keats, letter to his brother George and his wife, February–May 1819, in *Letters of John Keats*, ed. Robert Gittings (Oxford: Oxford University Press, 1970), 210–51.

73 According to Hillman, it was on account of the sessions of these Church Councils that 'the soul lost its dominion' and 'our anthropology, our idea of human nature, devolved from a tripartite cosmos of spirit, soul and body (or matter), to a dualism of spirit (or mind) and body (or matter)'. See Hillman's essay 'Peaks and Vales: The Soul/Spirit Distinction as Basis for the Differences between Psychotherapy and Spiritual Discipline' I. 'In Search of Soul', in *Puer Papers*, ed. James Hillman (Dallas: Spring Publications, 1979), 54–5.

74 '*The psyche does not exist without pathologizing* . . . The soul can exist without its therapists but not without its afflictions', Hillman, *Re-Visioning Psychology*, 70–1 (emphasis in original). Compare Julia Kristeva: 'without a bent for melancholia, there is no psyche, only a transition to action or play', *Black Sun: Depression and Melancholia*, trans. Leon S. Roudiez (New York: Columbia University Press, 1980), 4.

75 For Ibn 'Arabi's concept of the *barzakh*, see William Chittick, *The Sufi Path of Knowledge: Ibn al-'Arabi's Metaphysics of Imagination* (Albany: State University of New York Press, 1989), 14, 117–18. Perhaps the simplest definition Ibn 'Arabi gives is 'Any two adjacent things are in need of a *barzakh* which is neither the one nor the other but which possesses the power (*quwwa*)

of both', in *Futūhāt al-makkiyya* (The Meccan Openings) I 304.16, quoted in Chittick, *The Sufi Path of Knowledge*, 118.

76 In the *Sakuteiki*, perhaps the earliest gardening manual in existence anywhere, and written in Japan in the second half of the eleventh century, the gardener is urged to 'listen to the request of the stone'. This insight is particularly germane to the Japanese garden, since the *Sakuteiki* describes the practice of gardening as 'the art of setting stones', but it can also be readily extended to almost any component of a garden, including the lie of the land or a given spatial disposition. There is an excellent English edition of the *Sakuteiki*: *Sakuteiki: Visions of the Japanese Garden*, trans. Jiro Takei and Marc P. Keane, in the 'Tuttle Classics of Japanese Gardening' series (North Clarendon, VT: Tuttle Publishing, 2008).

77 See *Daodejing*, ch. 6: 'The valley spirit that doesn't die / we call the dark womb / the dark womb's mouth / we call the source of Heaven and Earth / as elusive as gossamer silk/ and yet it can't be exhausted', trans. Red Pine (Bill Porter), *Lao-Tzu's Taoteching, with selected commentaries from the past 2,000 years* (Port Townsend: Copper Canyon Press, 2009), 12–13.

78 A translation of the Heart Sutra is included in Thich Nhat Hahn, *Awakening of the Heart: Essential Buddhist sutras and commentaries* (Berkeley, CA: Parallax Press, 2012), 411–12.

79 The usual word for 'emptiness' in Buddhist texts is *śunyātā*: it applies to both phenomena and self. '*Anattā*' is a related term meaning more specifically 'non-self' in the human context or 'without lasting substance or essence'. Along with *aniccā* (impermanence) and *dukkha* (suffering or unease), *anattā* is one of the 'three marks of existence'. For a scholarly and sympathetic discussion of the doctrine of 'not-self', see Steven Collins, *Selfless Persons: Imagery and Thought in Theravāda Buddhism* (Cambridge: Cambridge University Press, 1982). A correlation between not-self and spatiality is drawn by the Japanese philosopher Kitaro Nishida (1870–1945) in giving priority to 'the place of *mu* (Nothingness)'. 'As people we are by nature spatial beings, and the essence of that nature is "*mu*". We use a variety of means to make ourselves aware of the fact: the Japanese garden offers a metaphor for exactly this pursuit.' Quoted by Lee Ufan in his essay 'The Gardens of Kyoto', in Sophie Walker, *The Japanese Garden* (New York: Phaidon Press, 2015), 69. The intuitive connection between space/spatiality and emptiness is frequently expressed in Buddhist and Daoist influenced art as well as in discourses on contemplative practices. The focus on spaciousness–emptiness is seen as assisting in the process of letting go of grasping mental habits. It also gives rise to such characteristic terms as *yohaku no bi* (the beauty of empty space) and *shin kū myō u* (from true emptiness the wondrous being appears).

80 The two inscriptions read *Humani nil a me alienum puto* (There is nothing human I count foreign to myself) and *Neque suspirationem dei compassibilem* (Nor the sigh of the compassionate God). The first is a quotation from Terence's play *Heauton Timorumenos* I.i.25, trans. A. J. Brothers as *The Self-Tormentor* (Aris & Phillips Classical Texts, 1988). The remark has been much admired and debated from Cicero and St Augustine through to Montaigne. The second inscription is inspired by Henry Corbin's discussion of the Sufi idea of the Creation as a Sigh of Compassion in *Creative Imagination*

in the Sufism of Ibn 'Arabi, trans. Ralph Manheim (Princeton: Princeton University Press, 1969), 112–20, 'The "Pathetic God"'.

81 Wen Zhengming (1470–1559) composed his album of thirty-one views of the Garden of Artless Administration (Zhuo Zheng Yuan) *c.* 1533. For a detailed discussion of three of these views, see Hongde Wu and Stanislaus Fung's 'Three settings from *Wen Hengshan's Album of Poems and Paintings of the Garden of Artless Administration*, 1533', Chapter 3 in this volume.

82 'The Thoughts of Afar Terrace' is the tenth of the views.

83 See note 20.

84 Keats reports to his brothers George and Tom in a letter of 21/27 (?) December 1817 that he came up with the phrase 'Negative Capability' in a spontaneous 'disquisition' he had with his friend Charles Wentworth Dilke as they walked home from attending a Christmas pantomime in December 1817. *Letters of John Keats*, 43.

PART IV

The garden retreat and mortality

9

Pope's Arcadia: Death re-enchanted

Yue Zhuang

Introduction

In a canonical article, Giorgio Melchiori underscores death as Alexander Pope's central preoccupation.[1] The pivotal role of the funereal '*Et in Arcadia Ego*' theme in Pope's poetic works ranges from the lamentation of Daphne's demise in the pastoral landscape to the poignant fate of Lodona, pursued by Pan, dissolving into a stream.[2] The frontispiece of Pope's mature work, *Essay on Man*, portrays the poet in serene repose as he leans against a sarcophagus amidst a sombre woodland, with a dilapidated Colosseum looming in the background (Figure 9.1).[3]

Born into a Catholic family during a Protestant regime, Pope's obsession with death may have resulted from the turbulent years following the revolution of 1688 and the early stage of the Hanoverian ascension, a time filled with tension surrounding the Jacobite risings. As Herman Feifel observes, consciousness of death, 'the shatterer of worlds', becomes more acute during periods of social disorganization.[4] Personal factors also weighed in. At the age of twelve, Pope contracted tuberculosis of the spine. This threatened his life at various times and seriously deformed his body. 'This long disease, my life'[5] cultivates his acute sensitivity towards death. Pope was preparing to die in 1705 at the precocious age of seventeen, when, he later told Joseph Spence, he was 'in so bad a state of health that . . . he resolved to give way to his distemper, and sat down calmly, in a full expectation of death in a short time'.[6] With such expectations, Pope embodied the Christian understanding that preparation for death is the proper business of a wise man. Among his boyhood reading was the famous book of devotion, Thomas à Kempis' *Imitatio Christi* (1475), which stressed 'to remember thine owne death in the death of Jesus, & make it your business to prepare your selfe to dye'.[7]

FIGURE 9.1 A painting probably by Pope: one version of the frontispiece to the *Essay on Man*. Courtesy of Bryn Mawr College, Special Collections.

Pope grew up in the woodland of Binfield, over 30 miles west of the City of London. His rural upbringing was a result of the 'Ten Miles ban' policy by William and Mary against Catholics. The young Pope, in his better days, seemed to have rejoiced in the 'Sylvan Scenes', where 'Descending Gods have found Elysium'. But death was never far away: 'in my Breast the Serpent Love abides'.[8] With all its vitality of life, the woodland landscape, for Pope, exemplified the enigmatic message in Nicolas Poussin's painting '*Et in Arcadia Ego*', which Ernst Panofsky famously interprets as 'Even in Arcady, here am I [Death]'.[9]

Contemplation of death was intrinsic to Pope's outlook and imagination, and necessarily became part of his day-to-day life. Writing to his advisor, Bishop Francis Atterbury in 1717, Pope confessed that the 'Contemplative life is not only my *scene*, but it is my habit too'.[10] This habit was continued at his new home at Twickenham in 1719, where he began to make a grotto and a garden. His poem to John Gay, a friend in the Scriblerus Club, captures his melancholic mood, as he imagines himself as a wounded deer dying in the tomb:

> What are the gay parterre, the chequer'd shade,
> The morning bower, the ev'ning colonnade,
> But soft recesses of uneasy minds,
> To sigh unheard in, to the passing winds?
> So the struck deer in some sequester'd part
> Lies down to die, the arrow at his heart;
> There, stretch'd unseen in coverts hid from day,
> Bleeds drop by drop, and pants his life away.[11]

Similarly, the drooping trees after a shower hold up their heads, yet to Pope, they 'remain checkd & witherd at the Root'.[12] While he seems to be delighted, for a moment, by the view that 'Our River glitters beneath an unclouded Sun' and 'Our gardens are offering their first Nosegays', he quickly checks himself, considering this a 'Pomp of style', and becomes preoccupied by a friend's death.[13]

If these daily reflections in his garden retreat were Pope's way of dwelling in his 'torn' world, such reflections have eluded significant scholarly attention.[14] The modern poet Sidney Keyes's note affords a hint: 'To the Middle Ages and the Elizabethans, Death was the leveler; to the seventeenth century, a metaphysical problem; to the eighteenth century, the end of life.'[15] This comment aligns with a prevalent thesis that characterizes the eighteenth century as an era of disenchantment, wherein death lost its spiritual significance, and the notion of leading one's life with a religious awareness of impending death became less widely accepted.[16] Influential factors such as the materialistic views on the soul earlier proposed by René Descartes and Thomas Hobbes, the scepticism regarding rewards and punishments in the afterlife articulated by Anthony Ashley-Cooper, 3rd Earl of Shaftesbury and Henry St John, 1st Viscount Bolingbroke, and the advancements in Newtonian science and medicine, all contributed to an Enlightenment mindset that secularized death, transforming it and life into distinct spheres of existence. This cultural shift was supported by the rise of polite society – emphasizing the avoidance of discomfort and the cultivation of agreeable emotions – underpinned by the emerging theology of latitudinarianism (deism), which actively embraced cheerfulness and advocated moderation in materialism.[17]

Typical of this deism, Shaftesbury advised that the 'contempt of death' should be regarded as 'the peculiar note of every generous and happy soul'.[18]

Yet as Melchiori and other recent scholarship have shown, this dismissal of the hereafter was by no means the only, or even the dominant, attitude in the early eighteenth century. Most Anglican priests, like George Berkeley and William Warburton, for example, considered it dangerous to remove the belief in the rewards and punishments of the afterlife, which was not only a deterrence but also a provider of hope in society.[19] Their argument was similar to that of their friend, Pope. Writing for the *Guardian* in 1713, Pope asked, 'what consolation is left to the Man who hath no Hope or Prospect of these things? . . . The *Free-thinkers* make it their Business to introduce Doubts, Perplexities and Despair into the Minds of Men, and, according to the Poet's Rule, are most justly punished by their own Schemes.'[20] According to Warburton, Pope once asked him to prepare an attack against Shaftesbury, considering that his '*Characteristicks* had done more harm to Revealed Religion in England than all the works of Infidelity put together'.[21]

Countering the free-thinkers' contempt for death and the dismissal of the final judgement, in Pope's and his contemporaries' neo-classical gardening, I argue, was a Christian-humanist approach that not only affirmed the concept of the immortality of the soul and the orthodox doctrine of reward and punishment, but also pursued less orthodox, but attractive visions of the beyond. Since neither the Old nor the New Testament describes the fate of the soul after death in great detail, Pope and his contemporaries freely borrowed from models of ancient literature such as the Elysian Fields and the Isles of the Blessed to imagine the journey of the soul after death and the green abode to which they hope to be admitted after death: shady grottoes and groves, luxuriant meadows, and fertile riverlands. This borrowing from ancient models did not separate the landscape elements from the philosophical or religious beliefs on the fate of the soul upon death and the funerary or initiatory rituals of the ancient mysteries, but reinserted them, via imagination, in day-to-day life. Such processes, which are usually overlooked by Pope scholars and garden historians, may be understood as means of re-enchantment. Countering the rationalizing attitude toward death, re-enchantment aimed to contain death in an overarching, existentially meaningful, ritual structure.[22]

Versed in the writings of antiquity, Pope, as a true heir of the English Renaissance, was committed to rejuvenating classical culture and making it relevant to contemporary Britain. The fate of the soul after death is a motif in both the *Iliad* and *Odyssey*, which Pope was translating from 1714 to 1726. While scholars have acknowledged the importance of landscape descriptions in Homer to Pope's notion of the picturesque,[23] what may be emphasized further is that Pope paid special attention to those verses on funerary rituals and necrotopography in Homer's landscape descriptions.[24] The Scriblerian friends praised Pope's translation of the *Iliad* for ensuring that the history of ancient rites was exact, a quality they found lacking in Thomas Tickell's translation of the first book of the same epic.[25]

Pope's approach to Homer, and antiquity in general, inherits the Hermetic-Neoplatonic tradition, which had been revived in the fifteenth-century Florentine Academy under the aegis of Marsilio Ficino and spread to Elizabethan England.[26] While this tradition was dwindling on the continent by the seventeenth century, it continued to develop in England as found among the influential Cambridge Platonists such as Henry More and John Smith. In this tradition, Homer was a prophet, who introduced the Egyptian Mysteries to Greece, and thereby developed the teaching of Plato and the interpretation of the later Neoplatonists in the third to fifth centuries. According to the Neoplatonists' exegesis of the *Odyssey*, Odysseus' voyage is an allegory of the death and new life of the soul, an idea that Pope took seriously. In his translation of Book XIII of the *Odyssey*, an account of Odysseus descending into Hades, Pope provided a detailed description of the Egyptian belief in the fate of the soul of the dead. He further compared Odysseus' descent to Virgil's imitation of the same in the *Aeneid*, Book VI: Aeneas' descent to the underworld, where he comes across both Hell and Elysium. Echoing the sentiments of the French Abbé Claude François Fraguier, Pope observed that Virgil imbues his narrative with a clearer sense of purpose, contrasting Homer's portrayal, which lacks distinction between the fates of the virtuous and the wicked: 'Homer gives a miserable account of a future state; there is not a person described in happiness . . . the good and the bad seem all in the same condition: whereas Virgil has an Hell for the wicked, and an Elysium for the just';[27] 'in the *Aeneid* the diseases, cares and terrours that Aeneas sees in his journey, are very happily imagined, as an introduction into the regions of death.'[28]

Pope's preference for Virgil's interpretation of the underworld consisting of a Hell for the wicked and an Elysium for the just suggests his acceptance of a tripartite scheme of the fate of the soul after death in the Hermetic-Neoplatonic tradition. In this tradition, all souls are pre-existent in the world. The body is thought of as a tomb or prison in which the soul is temporarily confined. Upon death, the soul will be freed from the body. Some souls are saved to Heaven, some are damned eternally in Hell. The majority of souls in between will go to the underworld, where they are purged of the taints of mortal passions and behaviour by wind or water or fire. They all pass through Elysium. A few virtuous souls remain there while purification is completed (before moving on to the ultimate Heaven after a long time), and the rest will be reborn on the earth as a punishment. These ideas were expressed in a range of Neoplatonist writings, and especially in Virgil's *Aeneid*, Book VI.[29] They continued to widely circulate in the early eighteenth century, as seen in, for example, John Dryden's translation of Virgil's *Aeneid* and André Dacier's *The Life of Pythagoras*.[30] Pope consulted and cited both authors for his translations of Homer.[31]

This tripartite scheme of the soul after death was compatible with the idea of inferno, purgatory and paradise as held by the Roman Catholic Church and it inspired Renaissance humanists' imagination of the necrotopography in both literary and spatial forms.[32] As a Catholic, there was no doubt that Pope firmly believed in the immortality of the soul, and Heaven and Hell.[33] For orthodox English Protestants following John Calvin, the idea of purification of the soul after death seemed suspicious, because Calvinists considered the state of souls was fixed at death. However, some Protestants questioned the static nature of the afterlife, including Francis van Helmont, Lady Conway and Henry Hallywell. Their theories allowed a Protestant version of purgatory in which the love of God provided an opportunity for mortal souls to be purified of their sins in order to be in blissful union with God.[34] As for metempsychosis, or transmigration of the soul, it is a settled notion in Pope, who considered it to be 'a very rational scheme' and 'would give the best account for some phenomena in the moral world'.[35] The Cambridge Platonists such as Henry More, Joseph Glanvill and George Rust all supported metempsychosis.[36] They believed that in the doctrine lay the key to vindicating the goodness of God in the face of evil and suffering in the world. But also it countered arguments by mortalists or annihilationists who argued that after death the soul died or slept between death and the day of judgement.[37] The scheme of the journey of the souls after death and they all pass Elysium retained its importance in Pope's time because it countered the deists' assertions against rewards and punishments and an increasingly disenchanted view of death.

For Pope and his like-minded contemporaries, the creation of landscapes of Elysium in their gardens in effect visualized the necrotopography through which the soul journeys after death and re-enchanted death. In Pope's garden, the imagined Elysium not only resonated with literary descriptions such as the descents of Odysseus and Aeneas to the underworld which were understood as allegories of the fate of the soul upon death, but also with the historical accounts of the initiatory rituals in the mysteries of antiquity, which represent the journey of the soul in the afterlife.

One such example of the ancient mysteries is the cult of Eleusis, a widely practised mystery dedicated to Ceres and Proserpina, goddesses of agriculture and death, in the Greco-Roman world.[38] Through the cult's complex rituals of initiation, the initiates were ensured immortality. The celebration of the mysteries involved purification in the Ilissos River and a long procession from Athens to Eleusis. There the procession descended into an underground temple, where the secret of the mysteries would be revealed. After re-emerging from the underground temple, the initiates found themselves in an Elysium-like grove or a charming meadow, where the initiation procession ended. Despite the secretive nature of mysteries, partial accounts of initiatory ceremonies were available in various ancient sources such as Pindar, Aristophanes and Apuleius, and inspired strong mystical and scholarly interest from the Renaissance onwards.[39] Pope obviously knew these ancients sources well and could have known the more recent work such as Joannes Meursius's *Eleusinia* (1619), followed closely by Antoine Banier's *The Mythology and*

Fables (1739–40).[40] The parallels between the ancient epic journeys and the initiatory routes in historical accounts are evident, because the epics usually contained initiatory scenarios. Early modern minds were well aware of these parallels. William Warburton famously argued that Aeneas' descent into the underworld was a coded description of the initiatory rituals of the Eleusinian mysteries.[41]

In his literary creations, Pope evoked these initiatory journeys (either literary or historical) as his ammunition to satirize Georgian society. For example, the gnome Umbriel's journey to the Cave of Spleen (*The Rape of the Lock*), a parody of Aeneas' descent into the underworld, allows Pope to dramatize contemporary debased ideas about Hell.[42] The dunces' procession in *The Dunciad* led from the City of London to Dulness' palace, alluding to the initiatory procession from Athens to Eleusis, characterized the hacks as worshippers of the false Apollo as they continually profaned the true.[43]

One may wonder why were Eleusinian mysteries important in the Age of Enlightenment. As Jan Assmann pointed out, initiation is a form of enlightenment.[44] For many of Pope's contemporary intellectuals such as John Locke and William Warburton, mankind needed revelation to gain a primordial understanding of religion and its true obligations.[45] The radical deistic concept of 'natural religion' undermined the essential role of revelation in religion as well as the aesthetic appreciation of nature. Pope rejected the radical deistic view by including Shaftesbury in his 1742 *Dunciad* among the dunces on the grounds that he had looked through nature not to nature's God but to 'that bright Image' 'Which Theocles in raptur'd vision saw, / While thro' Poetic scenes the Genius roves, / wild in Academic Groves'.[46]

Opposing the 'bright image' in the earl's 'academic groves', Pope's gardening, which is 'akin to philosophy',[47] is underlined by a deep awareness of the human limitation: death. 'That to study Philosophy', as Pope takes it from Montaigne, 'is to Learn to Dye'.[48] In this vein, one could posit that, for Pope, the garden or gardening itself, representing cyclical processes of generation and degeneration, or life and death as being inseparable, is a *memento mori*. A garden in which death is nowhere to be seen – like 'Timon's villa' in his *Epistle to Burlington* which he satirized – is downright 'Hell' for 'Ears Polite'.[49] Yet Pope's garden is more than a *memento mori*. Modelled upon the landscapes of necrotopography and sacred sites in classical and Renaissance literature, his Twickenham grotto and garden, as we shall see, enabled one to follow a path akin to the initiatory quests undertaken by epic heroes or initiands in mysteries. Infused with layers of mythological and mystic elements, Pope's Twickenham grotto and garden were crafted to evoke a sense of mystical initiation, wherein the secrets of the Eleusinian mysteries – the immortality of the soul and that gods are mortals – were simultaneously concealed and revealed. Serving as a sanctuary where the Christian concept of death is imbued with renewed mysticism, the grotto and garden emerge as spaces of re-enchantment and revelation.

The grotto

Laid out on five acres of rented land on the bank of the River Thames in Twickenham, Pope's villa no longer exists today, except for its grotto. One of the principal visual records of Pope's villa is a plan (Figure 9.2) drawn by his gardener, John Searle, in 1745, the year of Pope's death. The garden was separated from the house (2) by the road (4) from London to Hampton. To ensure private access from the riverside of the house to the garden, Pope dug a tunnel connecting the basement of the house at ground level to the garden. The tunnel and the basement were decorated as a grotto (3), which Jonathan Swift described as *ars poetica*.[50]

In modern scholars' readings, Pope's grotto is commonly taken as a nymphaeum – Pope happily discovered a spring while digging the tunnel – expressing the poet's nostalgia and passion for Italy.[51] What is less often noted is that nymphaeums originated from nymphs' caves. One such cave was Calypso's grotto, the subject matter of Book V of the *Odyssey*, where the nymph Calypso had Odysseus detained for seven years. The headpiece (Figure 9.3) of Book V of Pope's quarto *Odyssey* designed by William Kent, provides a visual impression of this grotto, a watery cavern dominated by groups of arches. Hermes, a figure seen as just arriving on the island, was to deliver a message from Jupiter requesting Odysseus' release. The visual resemblance between Pope's Twickenham grotto and Calypso's grotto has been noted by Maynard Mack.[52] Both grottoes are watery caverns with arches, and both are surrounded by a beautiful rural landscape of trees and meadows. In the centre of the image is a medallion, showing Odysseus

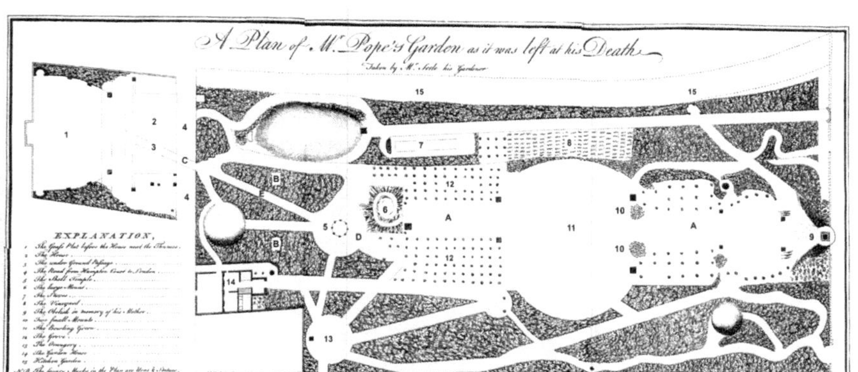

FIGURE 9.2 Pope's garden drawn by his gardener, John Searle, in his *Plan of Mr. Pope's Garden* (London: printed for R. Dodsley, 1745). Reproduced in *A Plan of Mr. Pope's garden: Gardens at Richmond, Kew, and environs*, ed. John Dixon Hunt (New York: Garland Publishing, 1982). Courtesy Forum Library, University of Exeter. For visibility, the reference numbers have been enlarged by the author. These further reference letters have been added: A. the Great Walk; B. the Wilderness; C. grotto exit; D. the glade; E. the sloping arcade.

FIGURE 9.3 Headpiece of Book V of Pope's *The Odyssey of Homer* (London: Printed for Bernard Lintot, 1725), designed by William Kent (1684–1748). Reproduced in Alexander Pope, *The Odyssey of Homer, Book I–XII*, in *The Twickenham Edition of Poems of Alexander Pope*, vol. IX, ed. Maynard Mack (London: Methuen & Co., 1967). Courtesy Forum Library, University of Exeter.

wielding an axe at trees, with a seated female figure looking on. The margin of the medallion bears an inscription: 'ΚΑΛΥΨΩ ΔΙΑ ΘΕΑΩΝ', which seems to be taken from line 11 in Book V which Pope translates as 'the magic bow'r'.[53]

The word 'calypso' in Greek also means 'concealment, or secret', a pun which Pope specifically pointed out, first in the opening of the *Odyssey* (Calypso as 'the Goddess of Secrecy')[54] and then in his notes to Book V.[55] The inscription in the medallion thus invites a reading of the image as an interplay between concealing and revealing. To Pope, revelation, the sense of mystery, protects truth from the profane with disguise. This notion is characteristic of Hermetic-Neoplatonic mysticism, in which the interest in myths, Plato's or otherwise, did not stem from a view of myth as a divine medium somehow above human reason. Instead, as Tae-Yeoun Keum explains, authors who participated in this mythic tradition were very much wedded to the centrality of reason in philosophy and politics and were interested in investigating how this distinct literary form could be deployed constructively toward that rational ideal.[56] They recognized that our reliance on certain deep, unconditioned imaginative frameworks in philosophical and political life was an inevitable consequence of the human condition and such frameworks would be neither discarded nor rationalized in the foreseeable future.[57]

Citing from the Greek Neoplatonic philosopher, Maximus Tyrius, Thomas Parnell, another of Pope's Scriblerian friends, explained in his essay on Homer that enriched the first volume of Pope's *Iliad*:

> It is the Madness of men to dis-esteem what is plain and admire what is hidden. This the Poets discovering invented the Fable for a Remedy, when they treated of holy Matters; which being more obscure than Conversation,

> and more clear than the Riddle, is a Mean between Knowledge and Ignorance; believ'd partly for being agreeable, and partly for being wonderful. Thus as Poets in Name, and Philosophers in Effect, they drew Mankind gradually to a search after Truth, when the Name of Philosopher would have been harsh and displeasing.[58]

Parnell's view fully resonated with discussions on matters such as the use of oracles in Pope's translation of the *Iliad*.[59] As Pope would have known, it was for this reason that George Berkeley condemned Shaftesbury for rejecting the Christian afterlife, because the beauty of virtue is not a sufficient principle or ground to engage sensual and worldly minded men in the practice of virtue.[60]

For Pope, the voyage of Odysseus is precisely such a 'disguise' deployed by Homer to guide people in the search for truth. Calypso's island, where Odysseus was imprisoned, may be a metaphor for the soul being confined by the body. Only by leaving bodily desire behind and setting off on a voyage to Ithaca did Odysseus' soul begin to seek freedom. In Book XIII of the *Odyssey*, when Odysseus finally reached the shores of Ithaca in the Phaeacians' boat, he was 'asleep' – a metaphor for the death of his body and the new life of the soul. As Pope expounded in a note:

> The whole voyage of Ulysses to his country, and indeed the whole *Odyssey*, has been turned into allegory: which I will lay before the reader as an instance of a trifling industry and strong imagination. Ulysses is in search of true felicity, the Ithaca and Penelope of Homer: he runs through many difficulties and dangers; this shews that happiness is not to be attained without labour and afflictions . . . and [he] is transported in his sleep to his country . . . his sleep means death, through which alone man arrives at eternal felicity.[61]

In particular, this note was made on the verses of landscape description of a quiet harbour sacred to Phorcys. In this harbour was hidden another cave of the nymphs, perhaps the most famous one in the Renaissance: the Naiads' grotto. In the widespread interpretation of Homer's description of the Naiads' grotto, *On the Cave of the Nymphs*, by the Roman Neoplatonist, Porphyry, the Naiads' grotto was endowed with cosmic significance; the cave was an image of the world, whereas the Naiads were the powers presiding over waters, on which the souls descend into genesis.[62] Porphyry's treatise enjoyed considerable popularity during the Renaissance and in the seventeenth century and was the inspiration for numerous Renaissance grottoes. Pope was familiar with Porphyry's exegesis, which he introduced in the notes:

> Porphyry (observes Eustathius) is of opinion, that the cave means the world; it is called gloomy, but agreeable, because it was made out of

> darkness, and afterwards set in this agreeable order by the hand of the Deity. It is consecrated to the nymphs; that is, it is destined to the habitation of spiritual substances united to the body . . .[63]

In Homer's *Odyssey*, it is mentioned that the Naiads' cave has two entrances. The one that looks north is the way down for men. The other, facing south, is for the gods.[64] In Pope's rendering of the *Odyssey*, it reads:

> Two marble doors unfold on either side;
> Sacred the south by which the Gods descend,
> But mortals enter at the northern end.[65]

With these two entrances, the cave was presented by Porphyry as a liminal realm where the soul transmigrates. As Pope's note on his translation continues:

> the two gates are the two *poles*; thro' the northern the souls descend from Heaven to animate the body, thro' the southern they ascend to Heaven, after they are separated from the body by death.[66]

The Twickenham grotto has similar features to the Naiads' cave. Not only did the grotto have a spring, which recalls Homer's image of 'Perpetual water o'er the pavement glide',[67] but also the grotto was a subterranean passage, with two entrances, one at each end. As Pope described in a letter to his friend and fellow Catholic, Edward Blount in 1725: 'There are connected to this Grotto by a narrower Passage two Porches', 'one toward the River, of smooth Stones, full of light and open; the other toward the Arch of Trees, rough with shells, flints, and iron ore.'[68] Might it be possible that Pope sees his own grotto, following the Naiads' cave in Porphyry's vision, as a liminal realm for him to practise the imagined journey of the soul after death?

Interestingly, having presented Porphyry's view on the transmigration of souls in his notes, Pope went on to suggest that he did not fully endorse the Neoplatonist interpretation, considering it to be too metaphysical and allegorical: 'I confess I should rather chuse to understand the description poetically, believing that Homer never dreamed of these matters, though the age in which he flourished was addicted to allegory.' [69] Pope's scepticism about Porphyry's metaphysical account bears the characteristics of British empiricism. As Jason Gulya suggests, individuals in the eighteenth century became increasingly sceptical of knowledge they could not verify with empirical data, personal experience or reasoning.[70] As a non-allegorical explanation, Pope introduced an account of the grotto as a place of Egyptian sacrifices, related to the religion of Osiris and Isis:

> they might carry the statues of the Gods in their procession thro' the southern gate, which might be consecrated to this use only . . . For that

> reason the Deities were said to enter: namely, by their images. As the other gate being allotted to common use, was said to be the passage for mortals.[71]

This note is certainly aligned with the statement in Parnell's essay on Homer, already mentioned: 'It appears indeed from [Homer's] Knowledge of the Ægytian Learning, that [Homer] was initiated into their Mysteries.'[72] Renaissance Hermeticism and Neoplatonism commonly recognized that it was Homer who introduced the Egyptian religion to Greece – its deities, doctrines and rituals. Preferring the historical and ritualistic account of the cave of the Naiads to Porphyry's metaphysical account, Pope was consistent in his overall approach affirming the role of ritual performances in sacred sites.

In his letter to Edward Blount, Pope wished to place a statue of a sleeping nymph in the grotto. 'It wants nothing to compleat it', 'but a good Statue with an Inscription, like that beautiful antique one which you know I am so fond of.'[73] He then set down four Latin verses, followed by his own translation:

> Nymph of the Grot, these sacred Springs I keep,
> And to the Murmur of these Waters sleep;
> Whoe'er thou art, ah gently tread the Cave,
> Ah Bathe in silence, or in silence lave.[74]

Pope did not specify the identity of the statue nor the authorship of the antique inscriptions. As Mack commented, the classical sculpture of a sleeping girl in combination with the anonymous inscription had captivated the Renaissance imagination since the fifteenth century.[75] A famous example, and one likely in Pope's mind, was the installation of the Cleopatra fountain in the Belvedere Statue Court at the Vatican in 1512.[76] Pope knew this well because, in his youthful age, he had translated a poem *On the Statue of Cleopatra* by the Italian humanist Baldessare Castiglione (1478–1529), dedicated to the statue. For Renaissance humanists, as for Pope, the identity of the sleeping girl with flowing water around her would be manifold, but the Neoplatonic belief in the new life of the rational soul upon death was always present.[77] Just as the voice of 'Cleopatra' praised Leo X for installing her statue as a fountain, Pope would have appreciated water for its generative power, as spelt out clearly in a note in the *Odyssey*: 'water is the first principle of generation'.[78] Pope's verse on the Cleopatra fountain ends with the following:

> Lull'd by these fountains the distrest may sleep;
> And while the Dogstar burns the thirsty field,
> These to the birds refreshing streams may yield;
> . . .

New greens shall spring, new flow'rs around me grow,
And on each tree the gold apples glow.[79]

Like the Latin verses he included in the letter to Blount, here in this poem the association of death (sleep), purification and generation is also indicated. The statue did not seem to have been installed. But since Pope desired to inscribe the verses in his grotto, of which the complex allusions to caves in the Neoplatonic tradition we have seen, it may be safe to suggest that the spring water is the genius of the place, and the grotto is a sanctuary where the soul purifies itself and prepares for the next stage: Elysium.

The garden – the wilderness

Just like those verses on the fountain of Cleopatra in Belvedere Court depicted a delicious garden, so there was a garden just outside the end of Pope's subterranean grotto. The transition from the grotto to the garden had particular importance for Pope, as he enthusiastically described in the 1725 letter to Blount:

> From the River *Thames*, you see thro' my Arch up a Walk of the Wilderness to a kind of open Temple, wholly compos'd of Shells in the Rustic Manner; and from that distance under the Temple you look down thro' a sloping Arcade of Trees, and see the Sails on the River passing suddenly and vanishing . . .[80]

Pope continued: 'You'll think I have been very Poetical in this Description, but it is pretty near the Truth.' Blount would have remembered Pope's confession in his earlier letter in 1721: 'I see one may go so far as to be Poetical, and hope to save one's Soul at the same time.'[81] The passage from the grotto to the garden, for Pope, would be akin to a spiritual journey of the soul, or a progress of initiation.

The gardening elements in Pope's garden resemble Italian Renaissance gardens: the grotto, the rustic temple, the arcade of trees and the mount. In Italian Renaissance gardens, visitors often have to access the garden through a grotto or a subterranean passage, thus experiencing a spatial transition between two different realms – the material and the intelligible, a peripatetic dramatization of the Hermetic-Neoplatonic world-view. A popular example would be Poliphilo's initiatory route described in *Hypnerotomachia Poliphili*, a hermetic prose romance attributed to the Venetian monk, Francesco Colonna, which was an inspiration for many Renaissance gardens.[82] Pope owned a copy of the 1600 Paris version of *Hypnerotomachia*.[83] Similar transitions from grotto to garden were practised in seventeenth-century English literary and artistic performances: poems, masques, theatre designs and gardens, all imbued with initiatory associations.[84] Pope would also know well the literary gardens in his favourite poet, Edmund Spenser's *Faerie Queene*, wherein both Proserpine's Cave and the Bower of Bliss in Book II (both indicators of deceit) were destroyed by Sir Guyon on his progression, and contrasted with the Garden of Adonis, the earthly paradise in Book III.[85]

As mentioned earlier, Pope's garden has long disappeared. The main visual evidence is the plan by Searle (Figure 9.2). Except for the area outside the grotto exit (C), the garden appears to be largely symmetrical, dominated by a 'Great Walk' (A–A), which forms an east–west axis. Outside the grotto exit is a freely planted grove, known as the 'Wilderness' (B). One can walk from the grotto exit to a glade (D) in the wilderness through a 'sloping Arcade' (E) formed by trees, terminating in a shell temple (5), erected in 1725, at the centre of the glade.[86] Next to the temple is a large mount (6), which Pope had started to raise around 1720, possibly with the soil he unearthed from digging the tunnel.

Pope left virtually no account of his garden, let alone any comment on its conception and possible associations. However, one may suggest that the garden he created as 'a place to stand'[87] would bear resemblance to the idealistic visions in his early poems such as *Pastorals*: 'Descending Gods have found Elysium here'. Like Virgil's Elysium, which resembled the sacred grove and meadow of Classical Greece and Rome,[88] the Twickenham garden featured forests and lawns. Both types were not uncommon among English gardens at the time,[89] although little has been explored in the context of ancient rituals and the fate of the soul of the dead. As the translator of Homer, where such descriptions were abundant, Pope was in a unique position to draw from these classical sources for his garden design and evidently did so, as Pierre Bonnechere demonstrates.[90]

Despite the lack of specific information about the nature of the trees in Pope's garden, we know from a Newcastle reporter's account that there were planted, apart from cypresses, lofty elms and other forest trees;[91] both elms and cypresses were used in antiquity for funeral pyres.[92] The hedges, such as the one at the banks of the 'Bowling Green' (11), typically consisted of laurel, bay and holly, all of which were commonly used in ancient rituals. All the vegetation was used adeptly to create confined spaces, as the said reporter recorded: 'Near the bounds of the garden, the trees united themselves more closely together, and cover the Hedges with a thick Shade', creating 'Wilderness Groves' and 'entangled' recesses.[93] Towards the middle of the garden, where the Great Walk opens up to the 'Bowling Green', there the hedges, combined with the 'wild and interwoven Branches of Trees', ensure the 'Privacy of the interior Parts'.[94] As Pope would have known, using vegetal regulation to create circumscribed spaces as sanctuaries was a practice in ancient Greece.[95]

Considering Figure 9.2 more closely, the journey through the garden may be divided into two parts, consisting of two distinctive experiences. The first part is essentially the wilderness, with a short walk from the exit of the grotto to the shell temple and the large mount (5–6) area. The second part is mainly regular, organized around an east–west axis starting from the large mount, through the Great Walk, which is flanked by the quincunxes (12); the axis extends to the west end of the garden, where originally there was a theatre, which was replaced in 1735 by the memorial to Pope's mother (9). In the first part, as previous scholars have noted, the approach by the 'sloping arcade' to the garden from the grotto exit (C) is at an odd angle to the main area dominated by the Great Walk.[96] This is because Pope's house was not lined up with the Great Walk, nor was the road perpendicular to the east–west axis, as Searle's plan, incorrectly, shows. Rather the house and the road faced northwest instead of due west, as James Sambrook's redrawing of Searle's plan suggests (Figure 9.4).[97] As a result, visitors emerging from the grotto exit and proceeding along the arcade thus found themselves entering the garden obliquely, not really aware of where they were with respect to the main area of the garden. The large mount (6), together with the two regular groves (12) seen from that point, further obscured the main axis to the west.[98] The spatial experience would have been both enchanting and bewildering, having something in common with the dialectical feelings of the initiand in the mysteries.

The second part of the journey through the garden is dominated by the Great Walk. Six stone urns (ordered in 1739 from Pope's friend, Ralph Allen) were placed along the axis, marking the entry to the end spaces and reinforcing the funereal nature of the progression. The axis is generally seen as 'the point of visual and emotional climax' in the garden by modern commentators.[99]

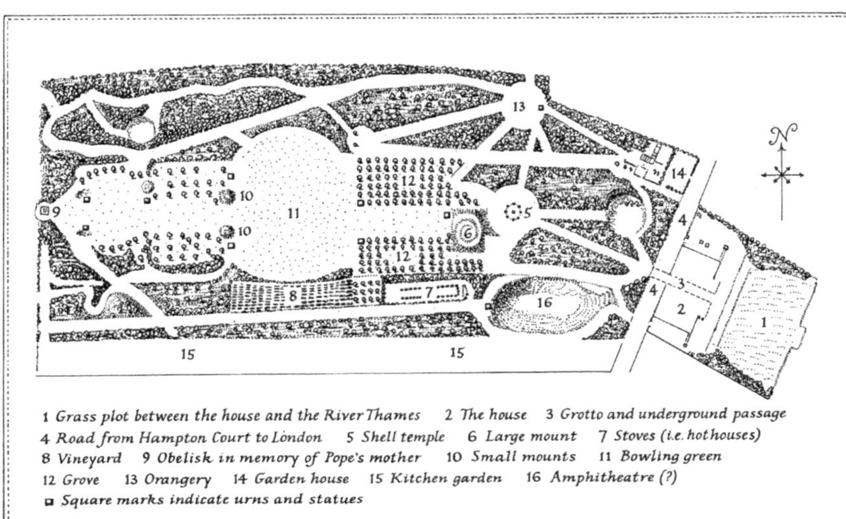

FIGURE 9.4 Redrawing of Searle's plan by James Sambrook (1972). The plan, in accordance with an added upright north arrow, appears to be reversed from Searle's plan. Reproduced in James Sambrook, 'Pope and the Visual Arts', in *Writers and their Background: Alexander Pope*, ed. Peter Dixon (London: Bell & Hyman, 1972). Courtesy of James Sambrook.

Furthermore, as has been noted by Michael Charlesworth, the geometrical layout of this part of the garden resembles a Palladian church.[100] Pope had also spoken of constructing a cathedral from trees: 'Here aged trees Cathedral walks compose'.[101] The spatial experience is solemn and grave. Moving along the path from the first part, the 'wilderness', to the second part, the 'church', might seem to indicate an evolutionary, linear vision of progress, from pagan mythology to Christianity, although the obelisk at the end of the progression would refer one back to Egyptian mysteries.

Parallels may be drawn between the first part of the garden and Pope's pastoral visions. For example, in his *Pastorals*, the shepherds and shepherdesses celebrated their love in Spring and Summer, before they mourn the death of Daphne in Winter through elegiac songs and rituals. In this aspect, Pope took inspiration from Edmund Spenser, whose landscape imagination in the *Faerie Queene* followed models of the sacred groves and meadows in the Greco-Roman world such as Venus' grove and the Garden of Adonis, where love, death and regeneration are the dominant theme. In line with the myth of Venus and Adonis, a closer model of the pastoral visions for Pope's Twickenham garden may be the myth of Galatea and Acis, which was elaborated in the musical opera, *Acis and Galatea*, first performed in 1718.[102] While the opera's lyrics were often attributed to John Gay, Pope contributed at least certain lines.[103] Based on the same myth, Pope had also written a poem 'Polyphemus and Acis' back in 1702.[104] The plot was simple, following the fable told in Ovid's *Metamorphoses*, that Acis and Galatea celebrated their love in the delicious pastoral landscape before a monster, Polypheme, out of jealousy, killed Acis with a rock. Galatea mourned Acis' death deeply. Reminded by the nymphs of her divine power, Galatea made Acis a deity, endowing him with immortality.

The opera opens with a joyous ambience that celebrates 'the pleasure of the plains' and the 'Happy Nymphs and happy Swains'. The nymph Galatea sings:

> Ye verdant Plains, and woody Mountains,
> Purling streams, and bubbling Fountains,
> . . .
> Too thin the shadow of the Grove,
> Too faint the Gales to cool my love.[105]

Her passion is echoed by Acis:

> Oh tell me, if you saw my Dear!
> Seeks she the Groves, or bathe in Crystal Fountains
> . . .
> Lo here, my love; turn, Galatea,
> Hither turn thy Eyes,
> See, at thy Feet the longing lies.[106]

This vivid scene seems to be visualized by William Kent in a drawing of the Twickenham garden (Plate 8). The painter's viewpoint may be near the large mount, looking eastwards towards the shell temple, placed in the centre of the drawing, and behind it, the grotto's entrance. To the left of the temple is a fountain in which a nymph is bathing. As no such fountain was found in Pope's garden, this scene is usually dismissed by scholars as a fantasy. An important exception is John Dixon Hunt, who noted that the drawing could be a gesture to the stage.[107] Pope, he proposed, could think of the garden in terms of theatrical scenes and in recognition of the long tradition of garden and theatre. And Kent, in particular, had been impressed by the theatrical scene of a fountain of Galatea in the famous Italian garden, Pratolino.[108] In light of the music opera, the dynamic between the young male behind and the bathing nymph in the front seems to represent that between Acis and Galatea, so vividly presented in the lyrics cited above. Their love is celebrated by the fecundity of nature – the verdant plain, the lush trees and bushes, the flowers and the bubbling fountain, all presented in Kent's drawing.

Yet as in all Neoplatonic pastoral landscapes, death is the condition of life. Thus is the beginning of the second part of *Acis and Galatea*, contributed by Pope:

> Wretched lovers, fate has past
> This sad Decree, no Joy shall last;
> Wretched Lovers, quit your Dream,
> Behold the Monster, Polypheme.[109]

In the music opera, Polypheme, who resides in a rocky palace, 'kills Acis with a great Stone which he gathers from a Rock'.[110] For a knowing visitor, would the large mount in Pope's garden be associated with the force of death and violence? As the Newcastle reporter describes, the large mount 'is composed of more rude and indigested materials; it is covered with Bushes and Trees of a wilder Growth, and more confused Order, rising as it were out of Clefts of Rocks, and Heaps of rugged and mossy Stones'. Among the mossy stones, 'a narrow intricate Path leads in an irregular Spiral to the Top'.[111] In Italian Renaissance and English gardens since medieval and Elizabethan times, the mount or mound was associated with death and burial customs.[112] Pope might also have in mind Leon Battista Alberti's commentary in *De re aedificatoria* that the ancients included funereal monuments in the gardens; the sepulchre of Erythraea was a great mound of earth planted with wild palms.[113] Pope would have had access to both the 1550 and 1565 editions of Alberti's treatise that were in the possession of his close friend, Richard Boyle, the 3rd Earl of Burlington and carefully studied by the earl.[114] (Inspired by Burlington, Pope embarked on a serious study of architecture from 1719 to 1723, while building his house and garden.[115]) The funereal nature of the mount is in accordance with that of the shell temple, a sacrificial structure associated with Venus, the goddess of regeneration.[116] Pope had owned a statue of Venus of Medici since at least 1719.[117] The statue was placed in his garden, along with a statue of Mercury (Hermes) as ornaments.[118]

The fact that the 'wilderness' is confined in the area along the walk between the grotto's exit and the shell temple, with the mount as a spatial defining element, is telling – a metaphor of life and death as a dialectic process. Alluding to the myth of Galatea and Acis, and indirectly, to Venus and Adonis, the garden of the 'wilderness' served to symbolize, and commemorate, the yearly decay and revival of life, especially vegetative life, which represents the cyclical processes of Nature. That the world mankind inhabits is a world of origination, death and regeneration had already been perpetuated by Spenser's verses on the death of Adonis in the *Faerie Queene*:

All be he subject to mortalitie,
Yet is eterne in mutabilitie,
And by succession made perpetuall,
Transformed oft, and chaunged diuerslie:[119]

Or, as Pope himself was to put it:

Builds Life on Death, on Change Duration founds,
And gives th' eternal wheels to know their rounds.[120]

The sensuous garden is not only filled with the impulses of life and its fertile seeds, but also with an allegory of the body passing on its way to the life–death cycle: 'all forms that perish other forms supply (by turns we catch the vital breath, and die)'.[121] This vision may be materialized as the view through one end of the tunnel. As Pope describes, 'from that distance under the Temple you look down thro' a sloping Arcade of Trees, and see the Sails on the River passing suddenly and vanishing'.[122]

For a pious Christian like Pope, the understanding of the fate of the mortal soul living and dying in such an ephemeral world, seen as part of the cyclical processes of metempsychosis, begets salvation. Pope's Elkanah Settle in the *Dunciad* parodies this when he asks, 'how long, thy transmigrating soul' has been traversing the world, and in what forms has it rolled 'from Boeotian to Boeotian'?[123] In the myth of Galatea and Acis, thanks to the nymph's divine power, the shepherd is made a deity. His blood is transformed into a fountain:

Purple be no more, thou Blood,
Glide thou like a crystal Flood.
Rock, thy hollow Womb disclose;
The bubling Fountain, lo, it flows;

Through the Plains he joys to rove,
Murm'ring still his gentle Love.[124]

Following this transformation, Acis reappears as an immortal. As Pope describes in his 1702 poem, 'Polyphemus and Acis':

With looks more azure, Acis might appear;
And Acis was; who now transform'd became
A crystal fountain, and preserv'd the name.[125]

With Acis' appearance, the fountain in Kent's drawing thus takes on another layer of meaning – immortality. It is apposite for Kent to juxtapose this fountain with a lively scene of ritual ceremony, as indicated by a smoking altar underneath the shell temple. Such a vision of the temple scene visualizes, for knowing readers, the vision of immortality, which Pope describes in his letter to Blount as a contrasting prospect to the sails 'passing suddenly and vanishing' on the river.

The Great Walk – the obelisk

When the visitors walked around the large mount (6), the spatial experience appeared to be both familiar and different. It was familiar in that it was another space dominated by trees and lawns. It was different because, in this part, the prevailing sense is order, measure and proportion. As already mentioned, the ground in this part of Pope's garden draws on the architecture of the Italian Renaissance, more specifically, a Palladian church.[126] This arrangement of having a more regular, or more sacred, part of the garden only becoming accessible after the visitor goes through the irregular part, might be read as an initiatory progression. Such an arrangement of an initiatory progression certainly has its precedents. In *Hypnerotomachia*, Poliphilo had gone through a series of initiatory scenarios, and only at the end of Book 1 became worthy of such a place as Venus' amphitheatre, where all the glories of Roman architecture are born anew.[127]

The dialectical economy between death and life, or between the passive and active energy, may be seen as a key message underpinning the Great Walk. From the relatively narrow walk at the beginning of the axis to the bowling green opening out in the middle of the garden, and to the space for a pause at the foot of the green bank upon which the obelisk stood, the visitor would have remained aware of the separate spaces: the successive compression and expansion of space along with the alternation of porous and hard edges through the use of widely spaced or densely clustered trees.[128] Horace Walpole's oft-quoted description of Pope's garden captured this dialectic aptly:

> The passing through from the gloom of the grotto to the opening day, the retiring and again assembling shades, the dusky groves, the larger lawn, and the solemnity of the termination at the cypresses that lead up to his mother' s tomb [obelisk] are managed with exquisite judgement.[129]

The cypress trees that Pope planted have been associated with ideas of death since antiquity. However, as Charlesworth has noted, because they are evergreens, they can also be associated with immortality.[130] Similarly, the obelisk, which Pope used to commemorate his mother, also has opposing significances. As a stony memorial, obelisks defy change, but they had begun, in ancient Egypt, by being connected with the sun because of their use as the gnomons of sundials. This would give them associations with the passing of time, mutability, and the inevitability of change and decay.[131] Since the obelisk can be glimpsed from afar soon after one walks around the large mount, it sets up in the minds of those initiated, from the beginning of the Great Walk in the garden, an expectation of Egyptian wisdom, the alleged origin of all religions in hermeticism – which the subsequent tour of the garden gradually unfolds.[132]

The inscription on the obelisk – '*Matrum optima*', the best of mothers – for Pope would signal both religious awe and filial piety, his Christian reverence for both his mother and Mother Nature. As Warburton argues in *Divine Legation*, the original religion in ancient Egypt was the monotheism of Nature, the archetype of 'natural religion', whereas polytheistic gods and goddesses were a product along with the creation of the state, as they provided people with moral and political orientation.[133] In this light, the image of Venus, the Great Mother of Nature and the goddess of regeneration, who is often seen as a variation of Isis and a conflation of both Ceres and Proserpina, is a disguise for the cryptic devotion for the Christian Blessed Mother.[134] In *Pastorals*, Pope evoked 'Celestial Venus', recalling the more esoteric meaning attached to her in the dialogues of Plato and developed by the Florentine humanists. In both *Windsor Forest* and the *Epistle to Burlington*, Pope gives Ceres overtones of the Christian Blessed Mother, especially by identifying her with the virgin Diana and the chaste Queen Anne.[135] Catholics were forbidden to construct places of worship, but here it seems Pope may have built his church to nature's Goddess, with a poetic disguise.[136]

Conclusion

The above reflections upon strolling through Pope's garden present the garden as a place for reimagining death and the mapping of its meaning. If we have unveiled an intricate web of symbolisms and allusions to myths and mythologies, the intention is not to impose a static allegorical reading on the garden. Rather, they were to reveal the dynamic relations between humanity with our mortality and nature. As Mircea Eliade explains from a Platonic perspective:

> Myths are the most general and effective means of awakening and maintaining consciousness of another world, a beyond, whether it be the divine world or the world of the Ancestors. This 'other world' represents a superhuman, 'transcendent' plane, the plane of absolute realities.[137]

Or, as Alasdair Forbes underlines in Chapter 8, the garden is a liminal realm, existing between realities and worlds. In a rationalizing society where nymphs were claimed to have departed, where mystical elements of thought were being displaced, and where death was to be despised and regarded as a taboo, it was imperative to re-enact the garden as a sacred, liminal realm, and to afford modern man with a vehicle to re-contain and re-enchant death. The modern notion that English landscape gardens, with an allegedly liberal aesthetic, are the origin of nineteenth-century landscape cemeteries may therefore be reversed.[138] Instead, in the case of Pope, it was the desire for the soul of the dead to be located in this spatio-temporal realm that inspired his garden.

Pope re-enchants death in what appears to our modern mindsets an unusual way. The naturalistic appearance of the garden, despite its long-claimed 'free natural Taste, and unaffected Simplicity',[139] seems to be a poetic veil that is necessary to protect the truth – that the soul is immortal and gods are but mortals – from the profane. A hint is left in the form of the famous Horatian maxim which Pope inscribed on a piece of marble over the garden entrance to the grotto, and which is signalled in Kent's 'fantasy' sketch (Plate 8):

> *Secretum iter et fallentis semita vitae.*[140]

Although usually translated as 'a secluded journey along the pathway of a life unnoticed', we may follow Malcolm Kelsall's suggestion to elaborate on Abraham Cowley's alternative translation: 'a secret journey through a strait road by an innocent deceiver'.[141] This would resonate with the similarly enigmatic message that Kent placed for the reader of Pope's *Odyssey*, 'ΚΑΛΥΨΩ ΔΙΑ ΘΕΑΩΝ', as we have seen. On both occasions, we are invited to read the garden as a sphere of hermetic truth, which can only be expressed negatively. The cloak of privacy and secrecy that his garden retreat afforded him allowed Pope to nurture and experience the truth as he saw it there and the truth, in turn, shaped and guided him in how he made his way in the fleeting, torn world.

Acknowledgements

I would like to thank Alasdair Forbes, Michael Charlesworth and Chris Ewers for their helpful comments and discussion on earlier versions of this chapter. Thanks are also due to Mark Dorrian for his initial inspiration for this study.

Notes

1 Giorgio Melchiori, 'Pope in Arcady: The theme of *Et in Arcadia Ego* in his Pastorals', *English Miscellany: A Symposium of History, Literature and the Arts*, 14 (1963): 88–92.

2 Alexander Pope, 'Winter', *Pastorals* (1709), in *The Twickenham Edition of the Poems of Alexander Pope*, ed. John Butt, vol. i, *Pastorals*, ed. Emile Audra and Aubrey Williams (London: Methuen & Co., 1961), 88–95. Henceforward *TE*. Pope, *Windsor-Forest* (1713), *TE*, vol. i, 168.

3 Pope, *Essay on Man* (1733–4), *TE*, vol. iii-i, *Essay on Man*, ed. Maynard Mack (1950; repr., London: Methuen & Co., 1958), plate facing page 11.

4 Herman Feifel, *New Meanings of Death* (New York: McGraw–Hill, 1977), 4.

5 Pope, *Epistle to Arbuthnot* (1735), l.132, *TE*, vol. iv, *Epistles to Several Persons*, ed. John Butt (1939; repr., London: Methuen & Co., 1953), 105.

6 Cited in Morris R. Brownell, '"Like Socrates": Pope's Art of Dying', *Studies in English Literature, 1500–1900*, 20, no. 3 (1980): 415. See Joseph Spence, *Observations, Anecdotes, and Characters of Books and Men. Collected from conversation*, ed. James M. Osborn (Oxford: Clarendon Press, 1966), no. 69, 30.

7 Brownell, 'Pope's Art of Dying', 415. Pope paraphrased it in a poem as a twelve-year old. *TE*, vol. vi, *Minor Poems*, ed. Norman Ault and John Butt (London: Methuen & Co., 1954), 5–6.

8 'Summer', *Pastorals*, ll.59–60, 68, *TE*, vol. i, 76–7. For an exploration of the theme of 'death in Arcady' in his *Pastorals* in a political context, see Pat Rogers, 'Pope in Arcadia: Pastoral and Its Dissolution', in *The Cambridge Companion to Alexander Pope*, ed. Pat Rogers (Cambridge: Cambridge University Press, 2007), 105–117.

9 See Erwin Panofsky, *Meaning in the Visual Arts: Papers in and on Art History* (Garden City, N.Y.: Doubleday, 1955), 310. Poussin painted twice the theme of *The Arcadian Shepherds*, first in about 1627 and later in 1637–8. Pope may well have seen the earlier version which is now in Chatsworth. See Melchiori, 'Pope in Arcady'.

10 Pope to Atterbury, 20 November 1717, Alexander Pope, *The Correspondence of Alexander Pope*, 5 vols, ed. George Wiley Sherburn (Oxford: Clarendon Press, 1956), vol. 1, 454.

11 'To Mr. Gay', *TE*, vol. vi, 225–6. The poem was later included in Pope's letter to Judith Cowper, 5 November 1722, *Correspondence* 2, 142.

12 Pope to Judith Cowper, 13 July 1723, *Correspondence* 2, 180.

13 Pope to Robert Digby, 1 May 1720, *Correspondence* 2, 44.

14 When death is concerned, existing scholarship mostly focuses on the obelisk to his mother erected in 1735. See note 99.

15 Cited in Maynard Mack, *Collected in Himself: Essays Critical, Biographical, and Bibliographical on Pope and Some of His Contemporaries* (Newark: University of Delaware Press, 1982), 77.

16 For both a summary and a challenge of the thesis, see Thomas Walter Laqueur, *The Work of the Dead: A Cultural History of Mortal Remains* (Princeton: Princeton University Press, 2015), 184–7.

17 For Hobbes' materialism, see Philip C. Almond, *Heaven and Hell in Enlightenment England* (Cambridge: Cambridge University Press, 1994), 47. For Shaftesbury's conception of disinterested virtue undermining hell as a deterrent in this life, and on the whole Christian afterlife, see D. P. Walker, *The Decline of Hell: Seventeenth-century Discussions of Eternal Torment* (London: Routledge & Kegan Paul, 1964), 168–177. On Bolingbroke's resignation towards the possibility of 'nonentity' after death, see Brownell, 'Pope's Art of Dying'. For a careful treatment of attitudes to death in the early eighteenth century, see Matthew Craske, *The Silent Rhetoric of the Body: A History of Monumental Sculpture and Commemorative Art in England, 1720–1770* (New Haven: Yale University Press, 2007), For a general introduction, see Clare Gittings and Peter C. Jupp, *Death in England: An Illustrated History* (New Brunswick, N.J.: Rutgers University Press, 2000),

18 Shaftesbury, *Characteristicks of Men, Manners, Opinions, Times* (1714), 3 vols. (London, 1732), I, 24, cited in Craske, *The Silent Rhetoric,* 58.

19 Walker, *The Decline of Hell*, 174–5.

20 Alexander Pope, *The Prose Works of Alexander Pope. Vol.1, The earlier works, 1711–1720*, ed. Norman Ault (Oxford: Basil Blackwell, 1936), 140.

21 See Warburton, *Letters from a Late Eminent Prelate to One of His Friends* (New York, 1809), 26, cited in Nancy K. Lawlor, 'Pope's *Essay on Man*: Oblique Light for a False Mirror', *Modern Language Quarterly*, 28, no. 3 (1967): 308.

22 This definition of re-enchantment follows the discussion by Raymond L. M. Lee, 'Modernity, Mortality and Re–Enchantment: The Death Taboo Revisited', *Sociology*, 42, no. 4 (2008): 745–59, which proposes that the public arena is not as bereft or disenchanted of magical or mystical currents and movements as Max Weber's image of modernity seemed to suggest. Among scholarships on gardens, Eric MacDonald, '*Hortus Incantans*: Gardening as an Art of Enchantment', in *Gardening – Philosophy for Everyone: Cultivating Wisdom*, ed. Dan O'Brien (Oxford: Blackwell, 2010), 121–34, and Robert Harrison, 'Homeless Gardens', in *The Re-Enchantment of the World: Secular Magic in a Rational Age*, ed. Joshua Landy and Michael Saler (Redwood City, CA: Stanford University Press, 2009), 72–80 also argue that gardens became means for countering the trend of Weberian disenchantment, although their case studies are modern and they do not make links to death. The use of the term 'enchantment' is not uncommon in the context of early eighteenth-century garden or landscape aesthetics. For instance, Joseph Addison wrote in 'The Pleasures of the Imagination': 'when we are in the *Metamorphoses*, we are walking on enchanted Ground, and see nothing but Scenes of Magick lying round us' (*The Spectator*, no. 417, 28 June 1712). Pope himself in a letter to Robert Digby in 1722 fancied himself at Lord Bathurst's 'enchanted forest' (*Correspondence* 2, 115). While these references to enchantment were noted by garden historians, they were often treated as expressions of liberty of the imagination, rather than in the sense of countering the Weberian disenchantment.

23 David Clark, 'Landscape painting effects in Pope's Homer', *Journal of Aesthetics and Art Criticism*, 22 (1963): 25–8; Morris Brownell, *Alexander Pope and the Arts of Georgian England* (Oxford: Clarendon Press, 1977), 51; John Dixon Hunt, *Gardens and the Picturesque: Studies in the History of Landscape Architecture* (Cambridge, MA: MIT Press, 1992), 107–8.

24 For example, see *TE*, vol. vii, *The Iliad, Books I–IX*, ed. Maynard Mack (London: Methuen & Co., 1967), Book I: 114–6 (on sacrifice); Book V: 283–4 (on protecting the dead body). *TE*, vol. viii, *The Iliad, Books X–XXIV*, ed. Maynard Mack (London: Methuen & Co., 1967), Book XXIII: 485–533 (on the funeral of Patroclus). *TE*, vol. ix, *The Odyssey of Homer, Books I–XII*, ed. Maynard Mack (London: Methuen & Co., 1967), Book X: 371–6 (on the landscape surroundings of Hades); Book XI: 377–426 (on the descent into Hell).

25 See, e.g., Thomas Parnell to Pope, 27 June 1715 and John Arbuthnot to Pope, 9 July 1715 in *Correspondence* 1, 299, 305. The research and discussion on ancient rituals in the notes were mainly contributed by William Broome and Elijah Fenton, Pope's assistants.

26 Hermeticism encompassed the body of arcane writing known as the *Corpus Hermeticum* which was attributed in the early Renaissance to the Egyptian priest Hermes Trismegistos, the Persian magus Zoroaster, Orpheus, Pythagoras, and others. The syncretism of Hermeticism and Neoplatonism was effected by Marsilio Ficino and Giovanni Pico della Mirandola in the Florentine Academy in the fifteenth century. Ficino translated into Latin not only all Plato's surviving works but also the *Corpus Hermeticum*, the works of Plotinus, and several works, mostly magical, of the later Neoplatonists – Porphyry, Iamblichus, and Proclus. The importance of Hermeticism and Neoplatonism in early modern Europe and England has been discussed by generations of scholarship, e.g., Frances Amelia Yates, *Giordano Bruno and the Hermetic tradition* (London: Routledge and Kegan Paul, 1964); Almond, *Heaven and Hell*; Glenn Alexander Magee (ed.), *The Cambridge Handbook of Western Mysticism and Esotericism* (Cambridge: Cambridge University Press, 2016). On Pope's reception of this tradition, see Douglas Brooks-Davies, *The Mercurian Monarch: Magical Politics from Spenser to Pope* (Manchester: Manchester University Press, 1983) and his *Pope's* Dunciad *and the Queen of Night: A Study in Emotional Jacobitism* (Manchester: Manchester University Press, 1985); David Fairer, 'Pope and the Elizabethans', in *The Cambridge Companion to Alexander Pope*, ed. Pat Rogers (Cambridge: Cambridge University Press, 2007), 89–104.

27 *TE*, vol. ix, 379.

28 Ibid., 378.

29 R. D. Williams, 'The Sixth Book of the *Aeneid*', *Greece & Rome*, 11, no. 1 (1964): 48–63.

30 See *Virgil's Aeneid*, trans. John Dryden with introduction and notes (1697; New York: P. F. Collier & Son, 1909), 234–237. André Dacier's *The Life of Pythagoras* (London, 1707), x, 29–30, 377.

31 For example, on verse 92, 'Forbid to pass th' irremeable Flood', in Book XXIII of the *Iliad*, the note describing the Egyptian concept of man that consisted of three parts (an intelligent Mind, a Vehicle for that Mind, and a Body) cited André Dacier's *The Life of Pythagoras*, 71. See, *TE*, vol. viii, 490–1.

32 E.g., Francesco Colonna, *Hypnerotomachia Poliphili* (Venice, 1499), see note 82.

33 Brownell, 'Pope's Art of Dying'.

34 For a detailed discussion of these schemes, see Almond, *Heaven and Hell*, 72–9.

35 Spence, *Anecdotes*, I, 239.

36 Almond, *Heaven and Hell*, 32–3.

37 Ibid.

38 For a canonical study on the Eleusinian mysteries, see Carl Kerényi, *Eleusis: Archetypal Image of Mother and Daughter* (London: Routledge & Kegan Paul, 1967).

39 E.g., in Pindar, *Fragments* 137: ''In speaking of the Eleusinian mysteries, Pindar adds': Blessed is he who sees them and goes beneath the earth; he knows the end of life and knows its Zeus-given beginning', *Pindar*, ed. and trans. William H. Race (Cambridge, MA: Harvard University Press, 1997), 385; Aristophanes' *Frogs* narrates Dionysus' (disguised as Heracles) travels to the underworld, in which the chorus and dances of Eleusinian initiates were described, see *Aristophanes*, ed. and trans. Jeffrey Henderson (Cambridge, MA: Harvard University Press, 1998), 70–77; Apuleius, *Metamorphoses*, Book 11 contains the most detailed description of the mysteries among ancient sources. See *Apuleius: Metamorphoses Books 7–11*, ed. and trans. J. Arthur Hanson (Cambridge, MA: Harvard University Press, 1989), 239–95. The Homeric *Hymn to Demeter*, with its information about the Eleusinian mysteries, was not available in its full form until nearly forty years after Pope's death.

40 Joannes Meursius, *Eleusinia, sive de Cereris Eleusinae sacro, ac festo* (Ex Officinâ Elzeviriana, 1619); Antoine Banier, *The Mythology and Fables of the Ancients . . . Translated from the original French* (London: A. Millar, 1739–40). Cited in Brooks-Davies, *Pope's* Dunciad, 135.

41 William Warburton, *The Divine Legation of Moses Demonstrated in Nine Books*, 2 vols (London, 1766), i. 133, 135, 144, 149, 182, 189, cited in B. W. Young, *Religion and Enlightenment in Eighteenth-century England: Theological Debate from Locke to Burke* (Oxford: Oxford University Press, 1998), 207.

42 Robert James Merrett, 'Death and Religion in *The Rape of the Lock*', *Mosaic: A Journal for the Interdisciplinary Study of Literature*, 15, no. 1 (1982): 34.

43 R. G. Peterson, 'Renaissance Classicism in Pope's *Dunciad*', *Studies in English Literature, 1500–1900*, 15, no. 3 (1975): 440.

44 Jan Assmann, *From Akhenaten to Moses: Ancient Egypt and Religious Change* (Cairo: American University in Cairo Press, 2014), 96.

45 David Jan Sorkin, *The Religious Enlightenment: Protestants, Jews, and Catholics from London to Vienna* (Princeton: Princeton University Press, 2011), 57.

46 This point is made in Katherine Myers, 'Shaftesbury, Pope, and Original Sacred Nature', *Garden History*, 38, no. 1 (2010): 3–19 (12), although I depart from her main point which continued the conventional grouping of Pope with Shaftesbury in regard to their deist outlook. While it may be true that both regard God as Nature and truth, they reveal polar opposites on the question of how to approach truth. While Shaftesbury considers reason to be the only means to pursue truth and rituals and revelation are superstitions, to Pope, the sense of mystery, like fables, protects truth from the profane by cloaking it in disguises. Pope could be called a deist, but he has to be an

'esoteric deist', to borrow Jan Assmann's term in his *Ancient Egypt and Religious Change*, 109.

47 Pope to Jonathan Swift, 25 March 1736, *Correspondence* 4, 6.

48 Montaigne, *Essaies* (1652), Book 1, ch. 19. Cited in Brownell, 'Pope's Art of Dying', 413.

49 See Aubrey L. Williams, 'A Hell for 'Ears Polite': Pope's *Epistle to Burlington*', *ELH*, 51, no. 3 (1984), 496.

50 Swift to Pope, 29 September 1725, *Correspondence* 2, 326.

51 Naomi Miller, *Heavenly Caves: Reflections on the Garden Grotto* (New York: G. Braziller, 1982), 77.

52 Mack, *The Garden and the City*, 51.

53 *TE*, vol. ix, 171.

54 Ibid., 27.

55 Ibid., 175.

56 Tae-Yeoun Keum, *Plato and the Mythic Tradition in Political Thought* (Cambridge, MA: Harvard University Press, 2020), 27.

57 Ibid.

58 Parnell, 'Essay on the life, writings and learning of Homer', *TE*, vol. vii, 70.

59 See, e.g., the note on verse 285 in Book xvi in *TE*, vol. vii, 250–2 contained an extensive quote from Temple Stanyan, *The Grecian History* (London: Jacob Jonson, 1707), 22–25.

60 Walker, *The Decline of Hell*, 176.

61 *TE*, vol. x, *The Odyssey of Homer, Books XIII–XXIV*, ed. Maynard Mack (London: Methuen & Co.), 7.

62 The Twickenham edition of the poems of Pope provided Porphyry's *De Antro Nympharum* (Rome, 1518) as the reference, see ibid. For the cited idea of Porphyry, see *De Antro* 10–19, in K. Nilüfer Akçay, *Porphyry's* On the Cave of the Nymphs *in Its Intellectual Context* (Leiden: Brill, 2019), 24.

63 *TE,* vol. x, 7.

64 Homer, *The Odyssey,* trans. E. V. Rieu, rev. trans. D. C. H. Rieu, introduction by Peter Jones (London: Penguin Books, 2003), 172.

65 Lines 133–35, *TE,* vol. x, pp. 8–9.

66 Ibid., 7–8.

67 Lines 122–125, 132, Book XIII, in ibid., 7–8.

68 Pope to Blount, 3 June 1725, *Correspondence* 2, 297.

69 *TE,* vol. x, 8.

70 Jason Gulya, *Allegory in Enlightenment: Literary Abominations* (Cham: Palgrave Macmillan, 2022), 4.

71 *TE*, vol. x, 9.

72 *TE*, vol. vii, 34.

73 *Correspondence* 2, 296.

74 Ibid.

75 Mack, *The Garden and the City*, 78. These purportedly ancient inscriptions, are now known to have been invented by a Roman humanist at the Vatican, Giovanni Antonio Campani, in a Greek style in the Florentine Academy in the 1460s. See Elisabeth B. MacDougall, 'The Sleeping Nymph: Origins of a Humanist Fountain Type', *The Art Bulletin*, 57, no. 3 (1975): 358–9.

76 Today's scholars identified that statue with Ariadne. For other statues of the same motif, see MacDougall, 'The Sleeping Nymph'.

77 For example, Pope's verses 'Or lull to rest the visionary maid' in *Eloisa and Abelard*, *TE*, vol. ii, 311, were derived from Ovid's description of Rhea Sylvia in the *Fasti*. For a discussion of many sleepers like Rhea Sylvia, Ariadne, and others in Greco-Roman art and their Neoplatonic meaning, see Sheila McNally, 'Ariadne and Others: Images of Sleep in Greek and Early Roman Art', *Classical Antiquity*, 4, no. 2 (1985): 152–92.

78 Note on verse 218, Book XIX, *TE*, vol. x, 204.

79 Pope, 'On the Statue of Cleopatra, made into a Fountain by Leo the Tenth. Translated from the Latin of Count Castiglione', *TE*, vol. vi, 68.

80 *Correspondence* 2, 296.

81 Pope to Blount, 3 October 1721, *Correspondence* 2, 87.

82 Francesco Colonna, *Hypnerotomachia Poliphili: The Strife of Love in a Dream*, trans. Joscelyn Godwin, 2nd ed. (London: Thames & Hudson, 2005). For an overview of such routes as initiatory processes in late Italian Renaissance and Baroque gardens, see Eugenio Battisti, *L'antirinascimento* (Milan: Feltrinelli, 1962), 134.

83 See Mack, 'A finding list of books surviving from Pope's library with a few that may not have survived', in *Collected in Himself*, 402.

84 The Caroline masques frequently evoked the cave/tomb in a pastoral setting, endowing a sense of mystic Arcadia to royalty, see, e.g., the design by Inigo Jones for Act v: Love's cabinet in the Neoplatonic pastoral *The Shepherd's Paradise*, in John Harris, et al., *The King's Arcadia: Inigo Jones and the Stuart court* (London: Arts Council of Great Britain, 1973), 163 (figure 302).

85 Pope identifies himself as a 'degenerate Ofspring' of Spenser, see *TE*, vol. i, 19. For a concise discussion of Pope's affinities with his predecessor, see Fairer, 'Pope and the Elizabethans'. Explorations of Spenser's literary gardens in relation to initiatory experiences in the Hermetic-Neoplatonic context is abundant, see, e.g. Humphrey Tonkin, 'Spenser's Garden of Adonis and Britomart's Quest', *PMLA*, 88, no. 3 (1973), 408, 411. James Nohrnberg, *The Analogy of The Faerie Queene* (Princeton: Princeton University Press, 1976).

86 The term 'Great Walk' was used by Pope in his letter to Lady Mary Wortley Montagu, 15 September 1721, *Correspondence* 2, 82. Cited in Peter Martin, *Pursuing Innocent Pleasures: The Gardening World of Alexander Pope* (Hamden: Archon Books, 1984), 46–60. The terms 'Wilderness' and 'sloping Arcade' are used in Pope's letter to Edward Blount, 3 June 1725, *Correspondence* 2, 296.

87 Mack, *The Garen and the City*, 232.

88 For the two types of ideal landscapes in Virgil's poems, see Ernst Robert Curtius, *European literature and the Latin Middle Ages* (London: Routledge & Kegan Paul, 1953), 192. On the use of sacred groves and meadows in ancient Greece for initiatory ceremonies, see Pierre Bonnechere, 'The Place of the Sacred Grove (*Alsos*) in the Mantic Ritual of Greece: The example of the *Alsos of Trophonios* at Lebadeia, in *Sacred Gardens and Landscapes: Ritual and Agency*, ed. Michel Conan (Washington, D.C.: Dumbarton Oaks, 2007), 17–42; Claude Calame, 'Gardens of love and meadows of the beyond: ritual encounters with the gods and poetical performances in ancient Greece', 43–56, the same volume.

89 Pope's garden could be one of those 'extended forest gardens', described in Stephen Switzer's *Ichnographia Rustica* (London, 1718). See Michael Charlesworth, 'Types of Gardens', in *A Cultural History of Gardens in the Age of Enlightenment*, ed. Stephen Bending (London: Bloomsbury, 2013), 55–6.

90 Pierre Bonnechere, 'Prairies et jardins grecs. De la Grèce de Platon à l'Angleterre d'Alexander Pope', in *Kêpoi: De la Religion à la Philosophie. Mélanges offerts à André Motte* (Liège: Presses Universitaires de Liège, 2001), Available online: <http://books.openedition.org/pulg/1086>. Accessed on 28 February 2024. Without referring to Virgil, Bonnechere points out the parallels between Pope's garden and those in ancient Greek sources like the *Odyssey*.

91 'An Epistolary Description of the Late Mr Pope's House and Gardens at Twickenham', in *The Newcastle General Magazine* (Newcastle-upon-Tyne, 1748): 25–28 (27).

92 Curtius, *European literature*, 192.

93 'An Epistolary Description', 27.

94 Ibid.

95 Calame, 'Gardens of love', 44.

96 James Sambrook, 'Shape and Size of Pope's Garden', *Eighteenth-century Studies*, 5 (Spring 1972): 450–55; Martin, *Pursuing Innocent Pleasures*, 49.

97 This fact was discovered by Sambrook. See his 'Shape and Size of Pope's Garden'.

98 Ibid.

99 E.g., Richard A. Etlin, *The Architecture of Death: The Transformation of the Cemetery in Eighteenth-century Paris* (Cambridge, MA: MIT Press, 1984), 173.

100 Michael Charlesworth, 'Alexander Pope's Garden at Twickenham: An Architectural Design Proposed', *The Journal of Garden History*, 7, no. 1 (1987): 58–72.

101 Spence, *Anecdotes*, no. 619, 256–7; the line is from Pope's 'Imitation of Cowley', *TE*, vol. vi, 47.

102 The most popular pastoral opera in early eighteenth-century England, *Acis and Galatea* is a collaboration between George Frideric Handel, John Gay and Alexander Pope. See Jeffrey Hopes, 'The Sounds of Early Eighteenth-

century Pastoral: Handel, Pope, Gay, and Hughes', *E-rea*, 14, no. 2 (2017), http://journals.openedition.org/erea/5741.

103 See editors' commentary in *TE*, vol. vi, 216–7.

104 Pope, 'Polyphemus and Acis', in *The Poems of Alexander Pope: A One Volume Edition of the Twickenham Text with Selected Annotations*, ed. John Butt (Bungay: Methuen & Co, 1985 [1963]), 18–22.

105 John Gay, Acis and Galatea, in *John Gay: Dramatic Works*, vol. 1, ed. John Fuller (Oxford: Clarendon Press, 1983), 268.

106 Ibid., 269.

107 Hunt, 'Landscape Architecture', in *William Kent: Designing Georgian Britain*, ed. Susan Weber (New York: Bard Graduate Center, 2013), ch. 14. Accessed 23 March, 2024. https://aaeportal.com/?id=-20580&cid=-20580.

108 See Hunt, 'Pope's Twickenham Revisited', *Eighteenth Century Life*, 8, no. 2 (1983): 31.

109 Gay, 'Acis and Galatea', 271; Pope, 'Lines from *Acis and Galatea*', *TE*, vol. vi, 216.

110 Gay, 'Acis and Galatea', 274.

111 'An Epistolary Description', 27.

112 For an example of the Snayl Mount in the Elvetham garden entertainments in 1591, see Amy L. Tigner, *Literature and the Renaissance Garden from Elizabeth I to Charles II: England's Paradise* (Burlington, VT: Ashgate, 2012), 59–60.

113 Claudia Lazzaro, *The Italian Renaissance Garden: From the Conventions of Planting, Design, and Ornament to the Grand Gardens of Sixteenth-Century Central Italy* (New Haven: Yale University Press, 1990), 57.

114 John Harris, *The Palladian Revival: Lord Burlington, His Villa and Garden at Chiswick* (New Haven: Yale University Press, 1994), 65.

115 Martin, *Pursuing Innocent Pleasures*, 41–3.

116 Venus as a goddess of regeneration is a central motif in Spenser's *Faerie Queene*, see Harry Berger, Jr. 'Spenser's Gardens of Adonis: Force and Form in the Renaissance Imagination', *University of Toronto Quarterly*, 30, no. 2 (1961): 136–9. Also representing motherly love, this side of her identity is linked to the heavenly Aphrodite, born in the sea. See Paul Friedrich, *The Meaning of Aphrodite* (Chicago: University of Chicago Press, 1978), 11, 45; David R. Coffin, 'Venus in the Eighteenth-century English Garden', *Garden History*, 28, no. 2 (Winter, 2000): 173–193. Kent's shell temple may be linked to Venus' temple, as described and illustrated in *Hypnerotomachia Poliphili*, see *Hypnerotomachia*, 205. Kent went on to design a Temple of Venus at Stowe in 1731. Kent's drawing of Pope's garden (Plate 8) also recalls the Roman motif of Venus' triumph (or birth in the sea).

117 Pope commissioned a painting of this from Sir Godfrey Kneller, see Pope's poem, 'To Sir Godfrey Kneller, On his painting for me the Statues of Apollo, Venus, and Hercules.' Written c.1719, *TE*, vol. vi, 213–4.

118 See Appendix B, 'The inventory of Pope's goods taken after his death', in Mack, *The Garden and the City*, 258.

119 Edmund Spenser, *The Faerie Queene, Books Three and Four*, ed. Dorothy Stephens (Cambridge: Hackett Publishing, 2006), 125 (Stanza 47 in Book 3).

120 *Epistle to Bathurst* (1733), *TE*, vol. III–ii, 107.

121 *Essay on Man*, III.17–20, *TE*, vol. III–i, 93–4. Cited in James William Johnson, 'What Was Neo-Classicism?', *Journal of British Studies*, 9, no. 1 (1969): 56.

122 *Correspondence* 2, 296.

123 Brooks-Davies, *Pope's* Dunciad, 59.

124 Gay, 'Acis and Galatea', 276.

125 Pope, 'Polyphemus and Acis', 22.

126 See note 100.

127 Terry Comito, *The Idea of the Garden in the Renaissance* (Hassocks: Harvester Press, 1979), 165.

128 Charlesworth, 'Pope's Garden', 63; Etlin, *The Architecture of Death*, 175ff.

129 Horace Walpole, *A History of the Modern Taste in Gardening*, ed. Isabel W. U. Chase (Princeton: Princeton University Press, 1943), 28–9.

130 Charlesworth, 'Pope's Garden', 63.

131 Ibid.

132 That the obelisk was funerary architecture for the Egyptians was amply clear to Pope. See Pope to Martha Blount, 22 June 1724, *Correspondence* 2, 239.

133 Warburton, *Divine Legation*, Book 2. Cited in Assmann, *Ancient Egypt and Religious Change*, 96–7.

134 See note 116. Brad Boyd noted the dense network of Britannia/Astraea/Venus/Ceres/Virgin Mary imagery and symbolism in John Dryden's post-1688 Virgil translations, which were well-known to Pope. See Boyd, 'Against Arcadia: English Mock-pastoral and mock-georgic, 1660–1740' (Berkeley: University of California, PhD Diss. 2013), 241–242.

135 Ibid.

136 Malcolm Kelsall, 'Landscapes and Estates', in *The Cambridge Companion to Alexander Pope*, ed. Pat Rogers (Cambridge: Cambridge University Press, 2007), 166.

137 Eliade, *Patterns in Comparative Religion* (Cleveland: Meridian, 1963), 429–30. Cited in Arthur Versluis, *Platonic Mysticism: Contemplative Science, Philosophy, Literature, and Art* (Albany, NY: State University of New York Press, 2017), 29.

138 Etlin, *The Architecture of Death*, 164.

139 'An Epistolary Description', 27.

140 Cited in Kelsall, 'Landscapes and Estates', 165. On the Horatian motto, see also Introduction to this volume.

141 This translation was kindly suggested to me by Michael Charlesworth. Cowley's original was 'a secret journey through a straight gate by an innocent deceiver', cited in Kelsall, 'Landscapes and Estates', 165.

10

'Each stone has a life to tell'

Derek Jarman's garden at Prospect Cottage

Michael Charlesworth

This chapter concerns the garden made by Derek Jarman (1942–94). He is famous as a film-maker, with an assured place in the history of cinema. He is also known for his garden (on a general, rather than scholarly level) and his Queer activism. In addition to these activities, he was a lifelong painter; a writer, not only of film scripts and soundtracks but also of a series of memoirs; and a stage and set designer from 1967. He was born and lived his whole life in England, with periods in Italy and Pakistan as a child.[1]

In May 1987 Jarman, who had been living as a Londoner since 1960, discovered a house he wanted to buy in the remote seaside hamlet of Dungeness, at the southernmost tip of the English county of Kent. With money left him by his father, who had died the previous year, he was able to buy the house, while retaining the small flat he occupied in London's Charing Cross Road for the rest of his life. He moved into Prospect Cottage, Dungeness, by August 1987. It was a strange place for a garden-lover like Jarman to choose: there was, for one thing, no soil to plant a garden in. The cottage stood on an ancient beach, part of a peninsula built up entirely of large pebbles, where a tenacious community of commercial but small-scale fishermen shared the space with a giant nuclear power station.

The 'torn'-ness of Jarman's world by May 1987 can hardly be exaggerated. Indeed, he wrote of it himself: 'My world is in fragments, smashed in pieces so fine I doubt I will ever re-assemble them.'[2] The reason for this destruction had manifested itself in December 1986, when Jarman received a medical diagnosis that he tested positive for the Human Immuno-deficiency Virus.

He was HIV+, and the virus opened the door for the illnesses associated with AIDS. In those days AIDS was not just occasionally, but inevitably fatal. Jarman now knew he was going to die of AIDS-related illness: the only question was how long it would take.

It was not, of course, only Jarman's life that was torn. The virus was engulfing his community of friends and acquaintances. The mysterious illness first emerged to more general concern by about 1983. Already, in 1985, the painter Mario Dubsky had died of AIDS; he had been a friend of Jarman's since the 1960s. By the time of Jarman's own death in February 1994, the number of his dead friends and acquaintances, several of them his former lovers, included Ian Charleson, Ken Hicks, David Dipnall, Robert Mapplethorpe, Howard Brookner, Karl Bowen and Paul Bettell. The serial killer Michele Lupo, another ex-lover of Jarman's, died of AIDS in prison a year after Jarman's death. Compounding the disaster of his generation was the wholly inadequate response by the British government to the disease. The government was administered at the time by Margaret Thatcher and the Conservative Party, which was, in any case, generally hostile to people of a same-sex sexual preference.

One of Jarman's responses to the situation was to become an articulate activist trying to benefit and protect his Queer community in the face of government neglect and the hostility of those in power (this joyless crew included, but was not limited to, Members of Parliament, senior policemen, newspaper proprietors, and certain people who alleged they were speaking for Christianity and on behalf of television viewers). And after a delay occasioned by coming to terms with the new soil-less environment, and lengthy commitments film-making and directing elsewhere, Jarman made a garden. The way I have put this, dividing his Queer activism from his gardening in separate sentences, suggests that the garden was simply a sanctum or refuge from an embattled world; but this view might perhaps be too simple. For instance, much of Jarman's film *The Garden* (1990) was filmed in and around Prospect Cottage, and in that film Jarman unleashed, on behalf of the Queer community, condemnations of legal oppression and the torment inflicted by media moguls, together with critiques of bad education, of Christianity, even of complacent gay assimilationists. The film is a 'dream allegory' of Jarman's situation and that of his community, in which his garden becomes figured as the centre of creativity, when we see him on-screen in the film dreaming, being ill (various sequences show him in a hospital bed), writing the film in the garden, and filming there.[3]

Jarman's garden is perhaps particularly appropriate for inclusion in the present volume because Jarman's last partner, Keith Collins, linked the garden's design to the fact that Jarman owned a copy of the first English-language edition of *The Secret of the Golden Flower: A Chinese Book of Life*, dating from 1932. This book consisted of Richard Wilhelm's German translation of an old Chinese text of esoteric knowledge, accompanied by an essay written by Wilhelm (1873–1930), and a Commentary by the Swiss psychologist Carl Jung (1875–1961).[4] Jarman had a professional background in design – for ballet, opera, theatre and film. He was thus used to creating a setting for at least one character to appear in. Collins wrote in 2013 of Jarman's design of his garden:

> Seen from above the garden forms a tetractys, dolmen and barrows at each intersection, precise, Pythagorean; at the centre the hearth of the cottage, at the apex a white dog rose, the first that Derek transplanted, protected with a staff and quoit of hagstone pebbles. Enshrouded in the garden plan are recreations of diagrams from Robert Fludd's *Tractatus de Geomantia* (1687) and mandalas from Wilhelm and Jung's *The Secret of the Golden Flower* (1929).[5]

By the time he wrote this, Collins had had nineteen years of experience caring for the cottage and garden – time to contemplate design arrangements at length. With respect to Wilhelm and Jung's book, he appears to be thinking of page 97. There, in his 'Commentary', Jung wrote about *mandalas*: 'For the most part, the *mandala* form is that of a flower, cross, or wheel, with a distinct tendency towards [the number] four as the basis of structure. (One is reminded of the *tetraktys*, the fundamental number in the Pythagorean system.)'

A mandala is generally definable as:

> a) A symbolic circular figure, usually with symmetrical divisions and figures of deities, etc., in the centre, used in Buddhism and other religions as a representation of the universe, and serving esp. as an object of meditation. (b) In Jungian psychology: an image or archetype . . . of a similar circle visualized in dreams, held to symbolize a striving for unity of self and completeness.

This quotation is from the *Oxford English Dictionary*. (It is interesting to see this source using the very text from Jung and Wilhelm for some of its evidential references.) In another work, Jung asserted that 'modern mandalas' have 'close parallels in ancient magic circles'.[6] (Jarman was in general deeply interested in Jung's thinking and owned several other works by the famous psychologist.)[7]

While Jung referred to 'modern' mandalas drawn by his European patients, the frontispiece of *The Secret of the Golden Flower* is from the Buddhist world: the reproduction of a Tibetan mandala (Figure 10.1).

Here we can see the classic form, based on the circle: circles inside squares inside circles; the number four figured in various ways. The patterns that Jarman created in stones and plants in his front garden employ the same alternating elements (Plate 9).

The question arises, then, what did Jarman think he was doing in making mandala-forms in his garden? In addition to *The Secret of the Golden*

FIGURE 10.1 Tibetan 'Vaj-ry' mandala as reproduced as Frontispiece to *The Secret of the Golden Flower*, trans. Richard Wilhelm. London: Kegan Paul, Trench, Turner & Co. Ltd., 1932. Photo: Author.

Flower, he owned a book by another author that also contains a description, albeit brief, of mandalas. Tenzin Gyatso, the Fourteenth Dalai Lama of Tibet, had received the Nobel Peace Prize in 1989, and his book *A Policy of Kindness* appeared in response in 1990.[8] In that year Jarman's front-garden mandalas were spruced up and improved with coloured stones from the nearby beach, and some of them enlarged.[9] In response to a direct question about what mandalas are, from his audience at a talk (transcribed in the book), the Dalai Lama replied: '*Mandala*, in general, means that which extracts the essence.' He lists several different types of mandala, and then specifies: 'the main meaning is for oneself to enter into the mandala and extract an essence in the sense of receiving blessing . . . Because one is gaining a blessing and thereupon developing realizations it is called an extraction . . . of something essential.'[10] Auspicious visualization can lead to a changed view of reality.

There is one other possible source of insight about mandalas within Jarman's experience that might have stimulated him in the making of his garden. In the early autumn of 1987 he had been to Japan to attend screenings of his films, and, Collins tells us, had also visited the Hokoku-Ji temple in Kamakura.[11] He could well have learned more about mandalas there. As one scholar of Japanese culture puts it, a mandala can be a cosmic diagram; or a 'Pilgrimage or Liminal' marker, offering the opportunity for 'mental journeys, for interior pilgrimages'; or representing the idea of a 'circle of acquaintances or friends' for contemplation.[12]

Carl Jung's interpretation of mandalas is thus only one among the possibilities known to Jarman, although he would have been particularly interested in Jung's long 'Commentary' that he contributed to *The Secret of the Golden Flower* simply because of his enthusiastic reading of other works by the Swiss psychologist. In Jung's view, 'the mandala, the protective circle' is 'the traditional antidote for chaotic states of mind'; and the making of one constitutes 'an attempt at self-healing on the part of nature, which does not spring from conscious reflection but from an instinctive impulse'.[13] Jung himself wrote: 'The 'squaring of the circle' . . . could even be called the *archetype of wholeness*.' Writing of making 'individual mandalas' (as opposed to 'ritual mandalas'), he says:

> Their object is the *self* in contradistinction to the *ego*, which is only the point of reference for consciousness, whereas the self comprises the totality of the psyche altogether, i.e., conscious *and* unconscious . . . images of this kind have under certain circumstances a considerable therapeutic effect on their authors.[14]

To make a mandala is 'not only a means of expression, but works an effect . . . It reacts upon its maker. Very ancient magical effects are associated with this symbol.' One of the functions of a mandala is 'to guard by apotropaeic means' the 'innermost personality' from 'deflections through external influences'.[15] And Jung stresses the honesty of the process of mandala-making, which encompasses 'the sphere of the chaotic personal unconscious, in which everything is found that one would willingly forget and that one would at all costs admit neither to oneself nor to another and would anyhow not take to be true'.[16]

The signs are that an immune-compromised Jarman envisaged the making of mandalas in his garden as protective, even strengthening. However, Jung writes that such therapeutic effect happens only when the mandala is made spontaneously, not when deliberately imitated. This is an aspect of instinct. Just as animals have instincts, so humans take refuge in instinctive symbolizations; hence the existence of archetypes, and of a 'collective' unconscious, which we can conceive of not as a stored metaphysical memory bank, but as a set of instinctual behaviours, moods and attitudes, partly, at least, derived from our culture.

It is quite possible that Jarman simply decided not to believe Jung's idea that the therapeutic value of individual mandalas only came if the making of them was spontaneous. In any case, such an idea is contradicted by the words of the Dalai Lama that are quoted above. If, on the other hand, Jarman did accept Jung's warning, this means that the mandalas he made were ritualistic, even if such rituals were psychological – private to himself. His stone patterns contain centres and circumferences: they are made with a number of upright beach stones, of flint, each having an individual natural form, forming a circle around a central upright stone. Jung writes on the centre and the circumference (as in a round dance): 'Psychologically this arrangement denotes a mandala and thereby a symbol of the self, on which are aligned not only the individual I, but at the same time many others of like mind or linked destiny.'[17] And in Japan, as we have seen, a mandala can represent the idea of a 'circle of friends and acquaintances' for contemplation. Meditating on this idea allows an additional resonance to the poem by Jarman that begins, 'I walk in this garden holding the hands of dead friends / Old age came quickly for my frosted generation' (the poem refers to all his gay friends who had already died of AIDS).[18]

Without wanting to imply, in Jarman's relation with the concepts of ancient wisdom set out in Wilhelm's book, any dogmatic belief (in the sense, for example, of Christian belief in the reality of its myth), I do want to suggest that they at least interested him as ideas; and that they did so at a crucial part of his life, when his mortality had just been authoritatively announced, or confirmed. Given that he was, in general, very interested in ideas of various kinds, particularly those to do with psychology, it is not unreasonable to think that he might have been moved to make some mandalas, in case he found that they helped him to come to terms with looming mortality. He wrote that he began planting the garden as 'a therapy and a pharmacopeia', implying that he hoped that it would give him strength, whether psychological or physical (or better still a combination of both), to resist his illness[19]. He made remarks that imply an optimism, or a denial, about his condition, telling viewers of the television documentary *Know What I Mean* that he felt perfectly well in 1988; and in 1992 writing of opportunistic infections: 'I feel that, if I can recover once, there's no reason why I can't recover again.'[20] If planting herbs might create a pharmacopeia, then making some mandalas might also contribute to physico-psychological strength.

With regard to the other source specified in Keith Collins's 'Epitaph' quoted above, the works of Robert Fludd (1574–1637), there is no doubt that Fludd's hermetic philosophy was in Jarman's mind at least after his reading of Frances Yates's book, *Theatre of the World* (1969). In his own memoir, *Dancing Ledge*, he tells us, in a passage bearing a date of November 1982, that he worked on paintings that year based on 'technical drawings of Robert Fludd, Athanasius Kircher and seventeenth-century hermeticism'.[21]

About the garden design, Collins specifies a particular source in Fludd: his discourse on geomancy (earth-magic). Dungeness is the site of a famous nuclear power station, and in his book *Derek Jarman's Garden*, Jarman refers to the idea that he might be a magician himself: 'At first, people thought I was building a garden for magical purposes – a white witch out to get the nuclear power station.' He immediately denies this belief, however: 'It did have magic – the magic of surprise, the treasure hunt.'[22] In the *Arena* television interview he gave (1991), he can be heard asserting that 'everything is magical. The world is magical.' We can see Jarman's good-humoured irony in these statements: wanting to discount the idea that he is a magician, yet asserting that magic exists (and therefore, in an unstated implication, the possibility that he was a magician.) Magic, for him, would have been the imagination and its psychological power. Part of that evidently included forming a layout for Prospect Cottage based on one of Fludd's geomantic figures, which shows the five points similar to the five on a dice, with an additional one outside the cluster of five. An aerial view of Prospect Cottage shows the cottage itself as the centre of a quincunx (five points; see Google Maps, 'Layers'). The shape is constituted by three circles of planting around posts to the northeast, northwest and southeast of the cottage, while to the southwest – where, for whatever reason, circles of planted gorse are difficult to sustain – stands, from time to time at least, a post on its own. To complete Fludd's diagram, we need to add, as Collins suggests, the first plant put in by Jarman: the dog rose.

Objections can be raised. Collins mentions the dog rose in relation to the tetractys, which is triangular, not a quincunx. Four features roughly aligned can be made out on the eastern edge of the garden in aerial photographs, but beyond that such imagery does not reveal much, and from aerial photographs at least, a tetractys cannot be made out. And there are three dog roses marked on Jarman's 1988 planting plan.[23] Two are off-centre; the other, while centred in relation to both a quincuncial and tetractial plan, was planted very close to the window in the centre of the back wall of the cottage, so would not qualify for either plan in terms of spatial layout or extent. But perhaps the original dog rose did not survive (Jarman had trouble sustaining his roses, and eventually gave up an idea he had for a layout based on nineteen varieties of rose.)

Despite these objections, there are features at the right distance from the house to complete Fludd's geomantic form. The purpose of geomancy being to make arrangements of the most auspicious kind for a house and setting, we can reflect for a moment on what Jarman did.

Both his geomantic figure and his mandalas offer protection. One could go so far as to specify particularly protection from the east (in Dungeness, the only public road runs to the east of Prospect Cottage). This is the direction linked in Richard Wilhelm's notes on the *I Ching* to 'that which greatly disturbs'.[24] Jung wrote of strengthening and protective effects of mandalas. As we noted, he writes of those his patients made as 'not only a means of expression, but works an effect . . . It reacts upon its maker. Very ancient magical effects are associated with this symbol.'

Jarman worked by sincerity mixed with irony. He loved humour, as all his recorded interviews show, and good humour, as stories about this gregarious man reveal. He planted herbs as a pharmacopoeia, to protect against disease. Pushed, he would have no doubt denied that he believed in such a magical mechanism. Yet at the same time, part of him would have had reservations about such a denial, and believed, or at least hoped, that it might work. At any rate, he felt the need to announce to the readers of his memoir *Modern Nature* (1991) that 'The garden as pharmacopoeia has failed'.[25] In the passage of writing which precedes this conclusion, Jarman printed one of his own poems about the garden, which characterizes the garden as commemorative, and includes the words: 'each stone has a life to tell'. The garden might have failed as a pharmacopoeia, but not as a design of forms for remembrance. And it has a more direct physical life that brings benefit: immediately after announcing its failure, Jarman writes: 'Yet there is a thrill watching the plants spring up that gives me hope.'

The dynamic, here, is that of contraries – not quite contradictions or antitheses, but more like William Blake's idea of contrary states of the human soul that can occur in response to the same stimulus[26]. 'Without contraries is no progression', wrote Blake.[27] One state felt by Jarman is depression. The contrary, for Jarman, to his depression is presented within the passage from which the above quotations come as the compensation of making a garden of commemoration and remembrance, and being able to write poetry about it; and the compensations of life: a congenial neighbour to talk to, the roses to water.

We should try to make some connections between the various forces discussed so far. The film *The Garden*, in showing Jarman sitting and writing the film in his garden, reveals the garden to be the centre of creativity. ('I should have been a gardener', said Jarman the painter, set designer, film-maker and writer.)[28] He sits in a chair made by a friend from recycled wood, in a garden defended by his auspicious geomantic spatial design derived from Robert Fludd, and his mandala-forms. Mandalas, as we have noted, can be, in Japan at least, opportunities for calling to memory a group of like-minded friends. The Dalai Lama states that we can extract 'blessing' from them. The blessing for Jarman must be the memory of his dear dead friends, and the sense that he has done something, however personal, to commemorate them.[29]

Putting these factors together, including the character of the film, *The Garden*, one inference must be that the garden as the seat of creativity inspired further creativity outside its own boundaries, and that this creativity involved action and protest in the face of the disease that had struck and killed his commemorated friends. Indeed his films from the 'Dungeness years', including *The Garden*, attempt to dramatize the struggle, tragedy, and the happiness, of Queer life. They include *War Requiem*, in which Jarman intended to use scenes evoking the First World War, and then borrowed film of twentieth-century warfare in general, to refer metaphorically to the death toll of young men from AIDS-related ailments; *Edward II*, in which demonstrators from the Queer organization Act Up become the ill-fated king's supporters; *Wittgenstein*, in which the famous philosopher finds it impossible to be happy about acting his sexual inclinations; and, triumphantly, *Blue*, the film that solves the problem about how to speak directly about HIV/AIDS, while impenitently celebrating gay sexual love.[30] In addition to the films, Jarman staged an installation at an art exhibition in Glasgow in 1989. This protested against another 'tearing' of the world: the notorious Clause 28, a 'deeply misguided' piece of central government legislation designed to control teaching in schools and the activities sponsored by local councils, in order to stifle any suggestion that a gay relationship was an acceptable form of domestic life.[31] The use of tar as a material in the Glasgow installation makes a small, but telling, connection with Dungeness, where, as with all seaside communities around the British coasts, tar was a commonly used substance (unseen lumps of which were prone to getting stuck to visitors' feet, shoes and even clothes). Finally, in addition to all these works, the books that Jarman wrote during these years, *Modern Nature, At Your Own Risk, Chroma, Derek Jarman's Garden, Smiling in Slow Motion,* all explicitly took up the cause of embattled Queer and Gay people.

There were more stones in Jarman's mandalas than the number of dead and dying friends. The argument being presented here is that the friends and acquaintances commemorated in Jarman's garden mandalas are remembered for their own sake, but also as standing for a larger Gay and Queer community still working and struggling in the world – as Jarman was himself. The commemorative garden features did not show evident signs of the struggle. Unlike Ian Hamilton Finlay's Little Sparta, *polemos* (conflict) was not a theme or subject of the garden. Jarman's features were products of, and strengthening innovations in, his creative imagination, which informed his work in other media, much (though not all) of which contains references to the garden. This is especially true of the published memoirs, the film *The Garden* and the soundtrack of his film *Blue*.

Jarman's journal, *Modern Nature,* becomes part of the process of bearing witness for his community. The diary format, no matter how much the actual writing might have been a representation (a work of art), strongly suggests a recording of the varying swings in mood – the shifts in *anima*, to put it in Jungian terms – out of which some peace of mind can come. And a recording of irresolutions, of decisions taken one day that are ignored the next, of happiness contradicted, misery annulled. The volume contains poems, prose-poetry, accounts of the herbs planted for the pharmacopoeia, a blasting of the film world of Soho as 'ridiculous', and harrowing accounts of falling ill.

In other words, *Modern Nature* conveys all the spontaneities of a diary. It is a book in which Jarman can be a witness and a spokesperson for his community, and can attack delinquent and even hostile government and parliamentary figures. There is also a strong element of memoir or reminiscence in the book, including celebrations of sexual activity and the more joyful experiences of Gay and Queer life. In these processes the garden has an important presence. The entries about the health uses and benefits of herbs testify to the importance of gardens in culture in the past, just as the discovery of opium poppies growing locally show their continued importance in that respect. The local opium poppies cannot be harvested, but opium, the true analgesic, is a substance we all need, at least from time to time. (This is commented on by Jarman in his late painting, *Morphine*, 1992, and on the last page of *Modern Nature*: 'The night's dreams, aided and abetted by morphine derivatives, grew increasingly menacing.'[32] The book does not end on an optimistic note: 'my appetite lost for recording and writing.') But the garden (in this case any garden) is also commemorated for its imaginative, psychological, spiritually transporting effect: 'The gardener digs in another time, without past or future, beginning or end . . . As you walk in the garden you pass into this time . . . Around you the landscape lies transfigured. Here is the Amen beyond the prayer.'[33] Any garden might convey this non-linear time, offering a respite from the determinisms of daily life's clock-time; but Jarman's, with its built-in mandalas, may concentrate this feeling by virtue of their surrounding presence. And Jarman asserts that the experience of the garden transfigures the landscape around. This is surely a key dynamic of the idea of a garden retreat. And being able to step out of normal time is not only transporting, but spiritually refreshing and strengthening for the gardener.

The other half of the argument presented here is therefore that the garden gave Jarman personal strength, at least in the form of happiness. This he makes clear in his writing; but to support that with non-verbal evidence we can turn to the paintings made in these years. Some of the paintings of the 1990s are large elaborate works of social protest made at a studio in London with no overt reference to Dungeness. More pertinent for the present argument is a group of small landscapes made in the confines of the painting room at Prospect Cottage: the *Dungeness landscapes*. In these paintings the depiction of objects in space is less important than the communication of colours and light seen, felt and relished. Jarman called these 'ecstatic landscapes'.[34] They are attractive through being heavily worked so that the paint is extensively scored and moulded into three dimensions. The working of the particularly thick applications of paint connotes a considerable exuberance, which is physical as well as emotional. We gain a sense of an excited physical motion, repeated for its own sake, striating the paint purposively, though not precisely: an accidental element persists, albeit small in character.

The other, and more dominant, deeply exuberant element in these paintings comes from the colours used, their saturation and brightness, and the boldness of their juxtapositions. Plate 10 reproduces a painting that shows perhaps the power station lit up by night, or perhaps a fire on the beach.

We can gain, from considering the application, for example, in the bottom left of the painting, of crimson red paint on top of the drying black, the impression that the process of painting still has great excitement for Jarman, despite the difficulties he encountered with his health in the last years.

My description of this painting indicates that the actual motif of the painting has very little individual importance. Its function is to provide a pretext for the colours applied and the working of the paint. Whether power station or fire, the motif also organizes colour arrangements of the whole. In addition to this comes a general function, that of indicating Jarman's happiness with the stimuli that Dungeness provided. Such a general feeling, exuding a considerable attraction, emanates from each of these landscapes.

His last, posthumously published journal, *Smiling in Slow Motion*, contains passages celebrating the colours of the landscape and garden. 'The sun sets', he wrote on 15 June 1992, 'and a large moon, burning orange in an indigo sky, ushers in the night.'[35] The first of his 'ecstatic landscapes' had been painted in June 1991, coinciding with a sight at the Royal Academy of works by 'Braque and Vlaminck my favourite fauves with violet and mauve skies. Back to the drawing table at Prospect.'[36] Three days later he wrote: 'The garden glows fire-bright in the wind, all scarlets and blues, like my paintings.' By the end of May the following year he reports: 'I dreamt all night of writing a book on colour, not scientific or in any way academic, free floating through the spectrum.' Perhaps partly in response to this dream, on 10 June he bought several books on colour.[37] His own book about colour, *Chroma*, was published in 1994. Perhaps the books all helped with the paintings too, as had the fauvist paintings, but it was Dungeness that kept providing the urgent and gleeful stimulus: 'The sun set behind the tower of Lydd church in a blaze of pink and purple, and the clouds turned to indigo in a full moon which sailed through them into the night.'[38]

A painting from 1992, *Prospect (Et in Arcadia Ego)*, presents a startling and uncompromising array of colours. The juxtaposition of a solid bar of saturated orange, with a violet-purple circle, agitated into three dimensions, dominates this small work. A major visual obstacle comes in the form of the black metal plaque adhering to the centre of the violet circle, and impressed with the words that form the painting's title. The question arises, if any of these colours relate to the objective world? The 'pink and purple' blaze of the sunset of 12 June gives us something to go on, as does this entry on 8 June: 'The sun is reflected a dazzling pale orange in the sea'; but in any case seeing is embodied. So although summer 1992 at times featured heat, blazing sunshine and consequent intense colouring of an alternately parched and wetted land, the stark blackness of the central plaque forebodes a sense of mortality – the inevitability of the end being not very far away. *Et in Arcadia Ego,* the phrase inscribed on a tomb in a famous painting by Nicholas Poussin, has an ambiguity: it can be taken easily to mean that the subject is in Arcadia, that mythic or archetypal beautiful land of happiness, where the inhabitants, simple herdsmen and shepherdesses, fall in love and don't need to work very hard. Erwin Panofsky's famous essay on the subject argues, however, that the inscribed words on the tomb in Poussin's painting (and the presence of a skull on the tomb in Il Guercino's earlier painting of the same subject) imply that the words are uttered by Death itself: 'I, too, am in Arcadia'.[39] Two such contrasting but viable readings accord well with Jarman's tendency to work by comparison, juxtaposition, metaphor and irony. On 7 June 1992 he had written: 'I sat by the front door wafted by the clove-scented pinks, it is an idyll: *et in Arcadia ego*. I am so in love with this place please God I see another year.'[40] Here the easier meaning seems to apply strongly, though it is immediately modified by Jarman reminding us of his awareness of his own looming mortality. Not so much Death present in Arcady, perhaps, as a dying man in Arcady.

The Dungeness landscape series therefore presents a few puzzles, while a sense of energy and joyful achievement is exuded by all of the paintings. They testify that Jarman had found the strength that he had hoped for when he bought Prospect Cottage. The strength was psychological more than physical.

What, we might ask, has he left to us? Jarman believed in the Renaissance theory of the *anima mundi* (soul of the world). He had first read about this in the books of the scholar Frances Yates, around the time he was designing the giant set for Ken Russell's film *The Devils* (1971) and preparing his own adaptation of Shakespeare's play *The Tempest*, as well as his unrealized script for a film about the Elizabethan magician, John Dee (1974–9).[41] In this view, things in the world – matter – become ensouled. The idea has its origin in ancient civilization (Egypt, Greece, Rome). In the ancient world, priests or adepts could call divine presences into objects, but in Renaissance times the soul descended into the world ultimately from God. Thus in Jarman's film *Caravaggio* (1986), the character Cardinal de Monte tells the young painter that 'Matter is in life, and ensouled'. All of this sounds rather far-fetched for a twentieth-century sensibility to believe, but Jarman had again encountered the idea that objects in the world are ensouled, in his reading of the works of psychology written by James Hillman, the post-Jungian psychologist and cultural commentator. Hillman's view was a secular version of the idea. Referring to degradation of city life and our built spaces, he wrote: 'All things show faces . . . As expressive forms, things speak; they show the shape they are in . . . An object bears witness to itself in the image it offers, and its depth lies in the complexities of this image.'[42] Jarman himself tells us that 'Objects are alive, living participants in the scene'.[43]

Jarman was able to connect these ideas about objects with the beliefs of the socialist founder of the Arts and Crafts movement, the poet, prose writer and craftsman William Morris. In an interview for the magazine *Square Peg*, Jarman said, 'I think it's terribly important to restore that kind of William Morris philosophy' and explained the philosophy as a belief that everything that has survived comes invested with all the thoughts that people had about it.[44] This is another secular version of the *anima mundi*.

But we need to ask, where do these connected ideas leave the garden? The novelist Anthony Powell believed that magic amounted to the activation of the will. He cites the French poet Arthur Rimbaud, and the French magician Éliphas Lévi, as having also believed so.[45] The archaeologist and philosopher R. G. Collingwood wrote that a marriage ceremony is magic, activating the will of the congregation on behalf of the welfare of the wedded couple.[46] Following the logic of the ideas and beliefs sketched above, we could conclude that Jarman – who can be heard in the BBC *Arena* programme made about him saying that 'Everything is magical. The world is magical' – has cast a magic spell, and vested the garden features with powers.[47] As an adept of memory he has called down into things in the world – unusually large or interesting stones gleaned from the beach -- the memory of his friends, and, of course, in turn the memory of himself, now dead as his friends were when he made the garden. Like Keith Collins before them, who died in 2018 after twenty-four years of looking after the garden that Jarman lived in for six-and-a-half years, the present keepers of the garden have a more than material responsibility. Material objects, to be sure, are highly important in perpetuating the magic that Jarman worked; but if everything comes down to us invested with all the thoughts that everyone to do with it felt, then that intangible element, fragile because of its investment in the merely material, is the heart and soul of the place.

Notes

1 For overall views of Jarman's life and work, see the detailed biography by Tony Peake, *Derek Jarman* (London: Little, Brown & Co., 1999); Michael Charlesworth, *Derek Jarman* (London: Reaktion Books, 2011); the exhibition book *Derek Jarman Protest!* (Dublin: Irish Museum of Modern Art and Thames and Hudson, 2019); and S. Farthing and E. Webb-Ingall (eds), *Derek Jarman's Sketchbooks* (London: Thames & Hudson, 2013). For his films, a selection of available works includes: Jim Ellis, *Derek Jarman's Angelic Conversations* (Minneapolis: University of Minnesota Press, 2009); Steven Dillon, *Derek Jarman and Lyric Film: The Mirror and the Sea* (Austin: University of Texas Press, 2004); Michael O'Pray, *Derek Jarman Dreams of England* (London, BFI, 1996); and Roland Wymer, *Derek Jarman* (Manchester: Manchester University Press, 2005). Robert Mills considers one theme in Jarman's work: *Derek Jarman's Medieval Modern* (Cambridge: D. S. Brewer, 2018). For the garden, see in particular Jarman's *Derek Jarman's Garden*, with photographs by Howard Sooley (London: Thames and Hudson, 1995);

the exhibition catalogue, *Derek Jarman My Garden's Boundaries are the Horizon* (Garden Museum, London, 2020); and Michael Charlesworth, *Derek Jarman's Visionary Arts: Exploring Land and Depth* (London: Bloomsbury Academic, 2024).

2 Jarman, *Kicking the Pricks*, 235.

3 Jarman is quoted describing the film as a 'dream allegory' in Jim Ellis, *Derek Jarman's Angelic Conversations*, 272, n. 39.

4 All of these texts were translated from the German by C. F. Baynes. The German edition had appeared in 1929. Jarman's copy survives in the bookcases of Prospect Cottage, Dungeness, Kent.

5 Collins, 'Epitaph', in *Derek Jarman's Sketchbooks*, 250–1.

6 Jung, *Psychology and Religion*, iv, 109 (1938), cited in the *Oxford English Dictionary*.

7 For the garden at Prospect Cottage and Jarman's connection of it to Avebury, see Michael Charlesworth, 'Derek Jarman's Garden at Prospect Cottage, Dungeness, and his Avebury Paintings', *Studies in the History of Gardens & Designed Landscapes*, 35, no. 2 (April–June 2015): 172–82. Jarman as a Jungian: Tony Peake, *Derek Jarman: A Biography* (1999; New York: the Overlook Press, 2000), 192, and various places in Jarman's own memoirs.

8 The Dalai Lama, *A Policy of Kindness,* ed. Sidney Piburn (Ithaca, New York: Snow Lion Publications, 1990).

9 Evidence substantiating these dates lies in photographs taken at the time by the author.

10 The Dalai Lama, 86.

11 *Sketchbooks*, 179.

12 Elizabeth ten Grotenhuis, *Japanese Mandalas: Representations of Sacred Geography* (Honolulu: University of Hawai'i Press, 1999), 2–4, 6.

13 Martin Brauen, *Mandala: Sacred Circle in Tibetan Buddhism* (Stuttgart and New York: Arnoldsche Art Publishers and the Rubin Museum of Art, 2009), 226.

14 Jung, extracts from *Mandala Symbolism* cited in Denise Patry Leidy and Robert Thurman, *Mandala: The Architecture of Enlightenment* (New York and Boston: Asia Society Galleries, Tibet House, and Shambala, 1998), 162.

15 Jung, 'Commentary', *The Secret of the Golden Flower*, 100.

16 Quoted by Brauen, *Mandala*, 230.

17 Ibid., 227.

18 The poem forms part of the soundtrack of Jarman's film *The Garden,* and is printed in Jarman, *Derek Jarman's Garden*, 81.

19 The quotation is from *Derek Jarman's Garden*, 12.

20 Jarman, *At Your Own Risk* (1992; Woodstock, New York: The Overlook Press, 1993), 122.

21 Jarman, *Dancing Ledge* (1984; New York: Overlook Press, 1993), 229.

22 *Derek Jarman's Garden*, 47.

23 This plan is in one of Jarman's notebooks and was consulted by me in Prospect Cottage in 2010.

24 Wilhelm, *The Secret of the Golden Flower*, 72.

25 Jarman, *Modern Nature: The Journals of Derek Jarman* (London: Century, 1991), 179 (Saturday 11 November 1989).

26 Blake uses the phrase 'the Two Contrary States of the Human Soul' in the titlepage of *Songs of Innocence and of Experience* (1789–94). The two sets of poems include many examples of differing responses to the same stimulus ('The Chimney Sweeper', for example).

27 William Blake, *The Marriage of Heaven and Hell* (1789), Plate 3.

28 Jarman's interview for Anthony Clare's radio programme, *In the Psychiatrist's Chair*, BBC Radio 4, 1990, published by Clare in book form under that title, 1992.

29 Robert Mills has a clear and persuasive view of the garden as commemorative in *Derek Jarman's Medieval Modern*, 125–30.

30 See in particular Jarman, *War Requiem: The Film* (London: Faber and Faber, 1989); and the voiceovers for *Blue,* printed in Jarman, *Blue* (New York: David Zwirner Books, 2023). All the films mentioned here except *Edward II* contain scenes of, or references to, gardening.

31 Karim Rehmani-White, in *Derek Jarman Protest*, 206.

32 *Modern Nature,* 314.

33 *Modern Nature*, 7 March 1989, 30. On the theme of time in Jarman's work, see Robert Mills, *Derek Jarman's Medieval Modern*; and Jim Ellis, *Derek Jarman's Angelic Conversations*, esp. ch. 6.

34 Jarman, *Smiling in Slow Motion*, ed. Keith Collins (London: Century, 2000), 32.

35 *Smiling in Slow Motion*, 149.

36 Ibid., 32.

37 Ibid., 134, 143.

38 Ibid., 144; 12 June 1992.

39 Erwin Panofsky, '*Et in Arcadia Ego*: Poussin and the Elegiac Tradition', in *Meaning in the Visual Arts* (Chicago: University of Chicago Press, 1955), 295–320.

40 *Smiling in Slow Motion*, 140.

41 Yates wrote about Robert Fludd; but for the *anima mundi*, see especially her *Giordano Bruno and the Hermetic Tradition* (Chicago: University of Chicago Press, 1964).

42 James Hillman, 'The Return of Soul to the World', *Spring: An Annual of Archetypal Psychology and Jungian Thought* (Dallas, Texas: Spring Publications, 1982), 71–93. Quotations 77–8. For the importance of Hillman's ideas to Jarman, see Charlesworth, *Derek Jarman's Visionary Arts*.

43 These comments, in the 1988 filmed interview 'Know What I Mean', specifically pertain to the design in *Caravaggio*.

44 BFI Jarman archives, DJ II, *Caravaggio* Box, item 7. Interview from 1986.

45 Powell, *Under Review: Further Writings on Writers 1946–1989* (London: Heinemann, 1991), 433–4 (review of Enid Starkie's book *Arthur Rimbaud*): '"magic" . . . is no more than the development of the will by the use of certain traditional disciplines', 434.

46 Collingwood, *Principles of Art* (Oxford: Clarendon Press, 1938), 66–7. Collingwood states that magic is designed to arouse certain emotions but not discharge them, thus developing or conserving 'morale'.

47 The programme was made by Mark Kidel and broadcast in 1991.

An afterword on withdrawal

Alasdair Forbes

Sometimes the best way to love the world is to leave it.[1]

Perhaps the most exalted idea we have about withdrawal can be found in the sixteenth-century rabbi, Isaac Luria's, term *tsimtsum*, which suggests that God's creation of the world only became possible through a prior contraction of His infinite nature, thereby permitting an empty space to emerge in which finite things could exist.[2] By its nature, the gesture of withdrawal is self-effacing and is often therefore written out of the account of the succession of conspicuous events that follow or replace it. Today, the self-effacing aspect of withdrawal, often characterized by the wish to go into hiding, is given little credit.[3] If withdrawal is worthy of notice at all, it is more likely to be accused of being self-obsessed. In fact, in today's world, the gesture of withdrawal is suspect. Forward-looking progress has no time for the step backwards or sideways or inwards, socio-economic necessity has no patience with seclusion. Nonetheless, to offset 'the frenzy and tumult of history', we have few better antidotes than withdrawal – both the act itself and the space within which it happens.[4] Nor does withdrawal have to be any more exclusively defensive for us than it was for God when He exercised *tsimtsum*. On the contrary, there is a strong case for saying that all real change begins with the act of withdrawal, that withdrawal and creativity are intimately linked. Defensive reflexes and transformative insights are as often paired as they are opposed. The change that emerges from withdrawal has many possibilities. Among them, to stay with Isaac Luria, would be the practice of *tikkun*, the desire to make good, or heal, or repair some broken aspect of our world which, notwithstanding conspicuous disasters, is always 'torn' in one way or another.[5] For Franz Rosenzweig, this act of repair was tantamount to an act of 'ensoulment' (*Beseelung*) of the creature by little acts of neighbour-love.[6] Might we not extend the possibility of such an ensoulment to any part of the created world?[7] In a garden especially, this notably delicate and humble act of repair can be carried out simultaneously and reciprocally on both place and person (most intimately between the devotedly tended plot of earth and the gardener themself). Principled withdrawal to a garden is a gentler practice than the radical path chosen by the 'martyrs' of reclusion or the path of the solitary hermit, being more disposed to a patient and flexible crafting of a way of life.[8] All the chapters in this book could be said to have woven their narratives around these themes of withdrawal and repair in one way or another. They address not just the virtue of withdrawal, but its conscience. In different contexts of necessity, from considered to urgent, withdrawal here is always a voluntary, and often a self-questioning, act.[9] And it is also always withdrawal to a garden. Why is that?[10]

First of all, we should recognize it is because the act of withdrawal always needs its concomitant spaces, be they caves, cells, huts, forests, islands, deserts or mountains. This partnership between withdrawal and its locales, the 'geographical imaginary' of reclusion, is very important.[11] As the seventeenth-century Japanese Zen monk Isshi Bunchu (1608–48) maintained, we always tend to underestimate the extent to which our physical surroundings determine the kind of practice we are able to do.[12] But why, then, the specific locale of a garden? This is surely because, for most of us, if we are lucky enough, the locale of a garden speaks to our very human need to live in a protective environment, which is nonetheless at the same time one porous to what is other than our humanly constructed world. Potentially at least, the human, that is the personal and the cultural, consorts with the natural and the preternatural, or, if we prefer, the cosmic and the metaphysical, in a garden, in the most poised and reflective ways. The garden, as these chapters confirm, has a unique topopoetic imaginary which lays down encouragements for how to dwell – that is, to live side by side with both the most intimate and most far-reaching horizons – within 'the very world which is the world of all of us', a world both reassuringly repetitive and inhospitably transient, a setting that is both our birthright, and yet unaccountable, nourishingly beautiful and irretrievably torn. '[13]

The more one familiarizes oneself with the cross-cultural 'patterns of disengagement'[14] which mark the life of withdrawal – ranging from 'substantive' to 'nominal' reclusion[15], from the homeless wandering of the ascetic or *sadhu*, through the monk in his cell and the solitary's hermitage, to Montaigne's library, Thoreau's cabin, Wittgenstein's and Heidegger's huts, all the way back again to the anonymity of the modern urban homeless hermit[16] – the more the image of a garden presents itself as potentially the ideal median setting and dwelling-place of reclusion. This ideal garden setting is aspirational and forgiving in equal measure, a friendly, close-to-hand environment in which to experiment with how to let illimitable horizons – the company of 'the permanent objects of nature' – inform our 'manners', our behaviour and our breathing, until they get indelibly under our skin.[17] The privileged intimacy of the garden setting allows such experiments to recur on an informal and daily basis. [18]

A garden is a part of the human world that offers itself discreetly to the natural world and, potentially, to other, more numinous worlds, too. Withdrawal to a garden shares with other modes of withdrawal a sense that the human world on its own – the world of our personal, working and social relationships – exhibits a narcissistic self-importance which, without appeal to extra-human perspectives, becomes illusory and claustrophobic. The apologists of withdrawal are very clear about this. In China, few cultural refrains run deeper than the one associated with the longing to escape '[falling into] their net of dust' (the pollution of the humanly constructed world).[19] While Confucius approved such a withdrawal in exceptional circumstances, he focused on the moral rectitude lying behind such a decision, and not on where such withdrawal took place (in garden, hut or countryside).[20] It is true there were Confucian writers who drew attention to the type of humble dwelling best suited to the retired life of the scholar, yet such concern was typically focused on the style of house or hut and how it was compatible with virtuous conduct; the whereabouts of these living quarters, be they garden or mountain retreat, received less attention. On the other hand, the celebration of the non-domestic appeal of the setting of withdrawal for its own sake gained traction through the 'cosmogonic turn' of Daoist and Buddhist thinking, haunting much of classical Chinese poetry, which faced towards the elemental setting of life, and was more inclined to turn away from placing humans (along with their rationality and their morality) at the centre of things. The garden, of course, is poised between these domestic and elemental directives. But insofar as the Daoist and Buddhist world-views engage a reality which insistently transcends morality rather than being framed by it, there is some justification in Li Chi's remark that part of the appeal of reclusion (in either a garden or a natural setting) was that it freed intellectuals and literati from 'the bondage of the moral world', thereby better answering 'the full demands of the mind' through encouraging participation in the alternative 'full life' provided by 'Taoist and Buddhist metaphysics'.[21]

In both China and Europe, traditional accounts of the garden retreat rely heavily on its virtuous ethos. This ethos, engrained in the philosophical accounts, characterizes both garden and gardener as 'happy, calm and unshaken',[22] possessed of a secure and confident stance and upholding a morally unequivocal position. However, when we get close up to gardens themselves, or indeed to the poets who are associated with them, pathos and ambiguity reveal themselves as closer to the constitution of both. For example, in the case of Tao Yuanming (365–427), who is often claimed as the archetypal exponent of serene reclusion in China, close reading of his work reveals a far more shadowed and complex character than his image in both popular and philosophical culture allows. In fact, Tao's verse, like his record of his life as a gardener/farmer, oscillates between contentment and dismay at his own circumstances and the ways of nature.[23] It is telling that during the Song dynasty (960–1279) scholars suppressed those textual variants in his work which threatened to destabilize Tao's image as a consistently affirmative poet.[24] This bias towards 'unshaken' stability as a virtuous ideal is a commonplace among morally minded scholars and philosophers, whereas in poets and gardens alike stability is always at a premium. The transfiguring moment of a garden's beauty is fugitive: awakening, transforming, but also ephemeral, a provisional epiphany of 'happiness' which quickly becomes no more than a promise, of course no less addictive for that.[25] The poetics of the garden knows this beauty, its pathos and its promise, which invests the complexion of the dwelling-place with sometimes delicate and sometimes scouring imprints of the torn world. For the garden is not only a retreat from the torn world, it is also its symptom. It is through the acknowledgement of this, through an acceptance that the garden includes disturbance and frailty within its physical constitution and expressive vocabulary, among other more vital, constant and irrepressible qualities, that it works its full power on us and becomes a companion not just to the mind, but to the heart and soul, of whoever withdraws there. But it is not only pathos that qualifies the ethos of a garden. It is also the garden's numinosity, its spirit of mystery, which undoes the 'courage of our convictions' again and again, in summoning us to, and effacing us before, the threshold of the sacred.[26]

In East and West alike, the case for withdrawal has always been disputed, not least by its most passionate spokesmen. Chrysippus already criticized Plato's, Aristotle's and Epicurus' advocacy of the life of withdrawal as the highest good for neglecting the more important social good of the *polis*.[27] Cowper and Wordsworth, two English poets whose central theme is the life of reclusion, persistently called into question the relative merits and demerits of withdrawal and sought to justify the individual's solitude in nature as the basis for a new kind of community.[28] The Trappist monk Thomas Merton and the Mahayana bodhisattva might concur that it is through the encounter with the deepest mystery of the self, seen against a background of cosmic transience, that the release of a transcendent compassion is more likely to occur. Du Fu, China's greatest poet, never ceased living out the dilemma of whether to serve or to live in seclusion, without ever resolving the issue. Unable to choose definitively between his 'yearning for palace gates', that is, to be of service at court and in government, and his 'wish for the gray isles', that is, to live as a Daoist recluse, he acknowledged sadly in his final poem that 'in taking care of my family and in the secrets of the cinnabar' (i.e. in fulfilling either Confucian or Daoist ideals) 'I have achieved nothing and my tears fall in a downpour'.[29] Nonetheless, the perspective of withdrawal, and the life of the hermit, has never – perhaps until now – ceased to haunt the imagination of the civilized.[30] G. K. Chesterton already noted that 'recent fashions are all for society without solitude' with the result that 'conventional sociability has become social suffocation'. He sought to remind us that the reason why a man became a hermit was 'because he was more of a human being, not less'.[31] And surely this is borne out by the unaccustomed range and depth of community the hermit is called upon to recognize as theirs, which is more remarkable for what it includes than for what it leaves out.

What are these alternative communities to which the individual in her or his solitude feels responsible? Traditionally at least, solitude was vital as the point of entry to them, whether in Rousseau, Wordsworth, Thoreau or Dickinson. The world that opened up to the solitary walker or resident consisted of 'the most sweet and tender, the most innocent and encouraging society [which] may be found in any natural object',[32] sometimes occasioning, as Dickinson put it, a 'transport of cordiality' with 'Nature's People'.[33] 'Landscape' itself for such a sensibility is much more than a scenic splendour, it is a 'society of living things'.[34] Or as Pater said of Wordsworth, 'to him every natural object seemed to possess more or less of a moral or spiritual life, to be capable of a companionship with man, full of expression, of inexplicable affinities and delicacies of intercourse'.[35] Wordsworth is unusual in the West in extending this 'primal sympathy'[36] to the widest horizons of the external universe, dovetailing, not always reassuringly, nature's 'huge and mighty forms that do not live / Like living men' with reciprocal 'mountings of the mind',[37] a process articulated with incomparable conviction through the rhythms and grandeur of his blank verse. In East Asia, the vital cosmic force of *qi* was, from time immemorial, similarly experienced as a psychophysical energy which pervaded the human and natural worlds, creating an indissoluble link between them.[38] So intrinsic was this experience to Chinese culture that veneration for the natural world of 'mountains and rivers' (*shanshui)* became a major subject of poetry and painting centuries before anything equivalent arose in Europe (which may partly explain why this theme of communion with nature, even when recognized as sublime, tends to be handled in a tone of quiet and poignant acceptance rather than with the enthusiasm of a personal revelation). It is worth pointing out that in China and Japan, as in Wordsworth, the alternative sense of community available extends not just beyond the anthropocentric world, but the biocentric one as well. This is evident, for instance, in a reverence for stone, variously termed 'petrophilia' or 'litholatry', and a perhaps even more privileged (and unusual) abiding sense of intimate companionship with spatiality. The Zen garden is entirely underwritten by such affiliations, referring back to Shinto rock worship and the Buddhist favouring of non-substantive 'emptiness' (śūnyatā); perhaps it is too seldom pointed out that space is our closest phenomenological echo of the spiritual experiences of nothing *(Jap. mu*) and emptiness.[39]

The intrinsic compatibility of solitude and withdrawal with, and preference for, this wider framework of communities than a conventional social conscience likes us to exercise is welcome and valuable in itself. It enlarges and refines our sense of what we belong to. Parts of our nature which stagnate in the addictive company of our fellow humans are rejuvenated and liberated, or even first discovered, when once we enter a world that exists beyond the endless cycle of societal distractions, when once we find ourselves entitled to dwell 'under the shade of the greenwood tree' or 'within hearing of the gibbon's cry'.[40] Trees, animals, stones, spaces also teach us who we are and distance us from who society wants us to be.

But how far then do reclusive 'patterns of disengagement' continue to reflect an 'involvement with mankind'?[41] If we listen to the voices of the recluses themselves, who often had scant regard for their own material welfare, the answer is, usually, a great deal. Even if, in some of the purest examples, no such conscious thought of justification for their way of life by way of assessing its impact on others occurs.[42]

So, for instance, Elzéard Bouffiier, the hero of Jean Giono's fable *The Man Who Planted Trees*, was not really thinking about improving the lot of his fellow citizens when he embarked on his vocation of planting trees in a desolate and sterile region of the Alpes-de-Haute-Provence. His concern was focused on the trees he planted and nurtured into maturity, against all the odds. Nonetheless, his selfless dedication to this project led, over decades, to the transformation of a barren landscape into a flourishing countryside where people felt once again encouraged to settle and build new villages, living out fulfilled lives without any inkling of the man whose labour had made all this possible. Through his initial calling to serve nature, Elzéard Bouffier thus unwittingly performed an anonymous service to mankind. He did not need to represent what he did as an ethical act in social terms. Living alone as a humble shepherd after the death of his wife and son, he simply did what he needed to do according to his vision of bringing back to life an abandoned tract of land, the portion of the earth allotted to him. He undertook an act of healing; it was his own practice of *tikkun*.

The account has the conviction and the simplicity of a fable. Follow your generosity wherever it leads. Do not prejudge what generosity entails. The solitude of the hermit gives of its heart freely with no expectation of reward or result. At its purest, the life of withdrawal is just that – a free gift, or the generation of those conditions of life which enable the solitary person to give of themself most freely in whichever direction their vocation dictates.

Elzéard Bouffier would hardly have written down his story himself. His heart was too simple, his inspiration too unselfconscious for that. But other notable solitaries such as Rousseau and Wordsworth have felt obliged to account for their reclusion in terms of its potential wider communal benefits. From their accounts it is self-evident that the goal of reclusion is not seen as a final retreat from the human world but as a strategy for how to reform it. So, for Rousseau, an abstention from complicity in the oppressive manners and devices of society leads, through intimate reacquaintance with the fundamentally benign world of nature, to a new form of 'social contract', while Wordsworth devotes an entire book of his major work, *The Prelude* (Book 8), to 'Love of Nature Leading to Love of Mankind'.[43] In Buddhism, the practice of 'detachment', involving a potentially tripartite physical withdrawal, mental withdrawal and withdrawal from the roots of suffering, so far from being an exercise in indifference or apathy, is wholly interdependent with a practice of both purifying and altruistic compassion (*karuna*), which characteristically embraces both the human and the non-human worlds.[44] Thomas Merton, writing from within the Christian tradition of monastic reclusion, echoes such a Buddhist perspective when he writes of the solitary's need to withdraw 'out of pity for the universe' and 'loyalty to mankind'.[45] It is wrong to assume detachment (or withdrawal) and compassion are at cross-purposes. The outwardly directed, clear-sighted compassion that extends to everything in existence is inseparable from the training of an attitude of mind 'to be radiated in meditation'[46] or, perhaps, from the cultivation of 'that serene and blessed mood' Wordsworth speaks of in his *Lines Composed . . . above Tintern Abbey*, which allows us 'to see into the life of things'.[47]

The individual gardens covered in this book were almost all private creations and were originally intended for private use. The exception is the park at Bry, which was apparently (re)designed, at least in part, for the use and benefit of the local townspeople.[48] The essentially private nature of the gardens does not prevent them addressing public concerns or adopting a counter-normative 'place to stand'. Maynard Mack, who formulated the idea of a poet's garden as a place to stand in relation to Pope's garden in Twickenham, suggests how such a garden can both be 'a rallying point for . . . personal values', which is comparable to the Chinese literati's practice of self-cultivation in their gardens, while also representing a cultural 'counter-order' to the official status quo of the day (in Pope's case the perceived corruptions of the court and the Walpole administration).[49] Pope's garden thus answered to the private needs of one who was 'design'd for solitude'[50] and served as his defensive base from which, as the satirist and man of 'affairs and fame',[51] to deliver increasingly stinging polemical broadsides, in superb verse, against the vices of the Establishment and contemporary decadence in general.[52] But even when a private garden is less overtly polemical than Pope's was (in one of its guises), its protest against prevailing cultural values can still be deeply felt, as many of the gardens in this volume show. The garden as a place of principled withdrawal cultivates a reflective horizon, poetic, philosophical and metaphysical by turn, which dissociates itself from the values of the 'Forum's madness'[53] or the 'net of dust' or, today, the dominant commercialized and socially acceptable convention of the garden. The impact of such a withdrawal on individuals, while discreet, may be far-reaching and even life-changing.

Because private gardens may express public concerns, they are not therefore 'altruistic' in conception any more than most other works of art are (or should be). However, the long-term fate of such gardens may prove to have an unsuspected public benefit, much as did Elzéard Bouffier's maturing forests. Unfortunately, Pope's, Qi Biaojia's and Sirén's gardens no longer survive. But when notable gardens do survive after their owners have died, they often take on a public afterlife, not possible or desirable during their creators' lifetimes. Examples here would be Wang Xianchen's garden in Suzhou and Derek Jarman's garden in Dungeness, which both now attract numerous visitors on a regular basis. Pope's grotto at Twickenham, the one part of his garden that has survived, has just been restored and is also now open to the public.

These gardens, then, were private domains where the garden-makers could feel secure in their withdrawal and free to 'follow their generosity' in creating whatever sort of garden they liked. That said, while the creative impulse's primary responsibility is to itself, almost no one makes a garden without the hope that it can be shared, even if only with like-minded friends.[54] One imagines some such company of 'the happy few' [55] always represents the kernel of who visits a garden on a regular basis during the owner's lifetime, radiating out into the audiences Qi Biaojia invited to his theatrical performances or the distinguished visitors Pope received at Twickenham from all over Europe, or perhaps, on occasion, to a wider public on 'open days'.

But, of course, the way in which a garden engages with a wider world than its physical boundaries describe is not restricted to the number and variety of its visitors. Prospect Cottage at Dungeness, Derek Jarman's garden, is especially defenceless in respect of having no real boundaries to protect it from either the natural elements or the gaze of strangers. More importantly, a garden, particularly a highly cultured one, can refer in all sorts of thematic and symbolic ways to worlds outside its walls. This is well brought out in Hongde Wu and Stanislaus Fung's Chapter 3 on the Garden of Artless Administration: Wen Zhengming's paintings of, and poems about, the garden make clear that features within the garden are intended to act as an imaginal trigger for all kinds of allusive and exophoric references. The vast and lonely scenes depicted (it is rare for more than one figure to be shown, and never more than two or three) conjure up a vision of remote mountains, quite unlike the actual setting of the garden in suburban Suzhou, where the visitor to the garden is encouraged to dwell in their imagination as if they were a wilderness recluse. However well-defined the physical boundaries of a garden are – and we should not underestimate the importance of this first line of defence which sets the precious world of the garden apart from the profane outside world – the life of the imagination within the garden knows no bounds. If the garden's walls offer a certain kind of protection from harsh weather and other unwelcome disturbances – a privileged 'encamping of the mind', as the poet Andrew Marvell puts it – this security guarantee they provide for the life of withdrawal at one level only serves to provoke the opportunity for a fully uninhibited play of the imagination on another.[56] At Little Sparta, Ian Hamilton Finlay's garden in Scotland, for instance, the garden is conceived as if it were a garrison against an inimical outside world. This initial stance of defiant and estranged withdrawal is then shown to guard an inner place which revels in drawing all kinds of subtle and provocative connections with both the natural and cultural worlds. A variant example of this strategy within the book is given by the Hermes courtyard at Plaz Metaxu. Memory provides another means for transcending the physical and temporal bounds of the actual setting of the garden, as exemplified in Osvald Sirén's retreat (the subject of Minna Törmä's Chapter 5), which was inspired, with the aid of many photographs, by his extensive earlier travels in China and Japan. Imagination and memory are also the dominant players in Iván Szántó's account (Chapter 4) of the gardens made across Europe and Asia in the image of Omar Khayyam's verse.

The garden we withdraw to can therefore be understood as both 'encamping' the mind and enfranchising it. The premise of security is almost always a given (the exception being Jarman's garden), but this often acts as a springboard for drawing imaginative connections between the garden and the real or metaphorical worlds beyond its walls. This expansion and retraction of the garden's sphere of influence ('capturing alive' distant remembered or metaphorical images) generally has a liberating and enriching effect on the garden, multiplying its references and layering its meanings. But not all breaches (or over-leapings) of the garden's defences are this benign, though. Nature itself has its destructive forces from which no garden is entirely immune. Disturbing thoughts have no greater respect for boundaries than exhilarating ones do. No one who makes a life work of their garden, or dedicates themself to it unreservedly, can sustain the illusion that either their garden or themself lives in isolation from the 'torn world'. Whatever external factors trigger withdrawal – political corruption or collapse in the case of Wang Xianchen and Qi Biaojia, the outbreak of war for Osvald Sirén, a terminal diagnosis for Derek Jarman, to mention only the most extreme examples in the book – once the garden becomes a reclusive dwelling-place, these conspicuous examples of a torn world are soon supplemented or replaced by others of a less importunate, but more insidious and abiding nature, relating to a wider world than the human one, on the one hand, but also to a more private world than the public one, on the other.

Abraham Cowley gives us a candid example of someone who disappointed himself in failing to live up to, or sustain, the life of innocence and equanimity assumed to be the hallmarks of the virtuous recluse.[57] For Cowley, who retired late in life to a rented estate at Chertsey in Surrey so as to fulfil his lifelong ambition of being 'master at last of a small house and large garden', found in practice that the blessing of the garden given to man by God 'As the chief help and joy of human life' turned out to be less exempted from vicissitudes than he had hoped: once having made the 'first and hardest step' of 'abandoning . . . this world, and . . . retiring from the noise of all business and almost company', he soon discovered himself 'disappointed . . . of that felicity' by 'several accidents of my ill fortune' which he was never able to overcome.[58] Cowley's admission that withdrawal to a garden, while removing one set of annoyances, quickly gave rise to others reads like a test case of Ian Kidd's contention that the garden, as a scene of innocence lost, cannot seriously hope to exclude moral or other harms from within its walls.

Yet it is a feature of the life of withdrawal that it practises turning such harms to advantage. For instance, one thought which permeates the life of withdrawal, and most of the gardens under discussion in this book, is the contemplation of death, for which, notoriously, there is little time or appetite in the manoeuvrings of the active life. Death has been called 'the mother of beauty', but it is also the conscience of withdrawal.[59] In one sense, death is the intimate companion of those 'glooms and solitudes' which, for Finlay, characterize 'superior gardens', because they are the 'worlds' of 'the proper work of an *exile*' (and 'Death is the extremest *exile*'), not mere hedonistic retreats.[60] But more subtly, perhaps, as in Daoism, death and life can be experienced as reciprocal tendencies, with the aptitude for withdrawal understood to be moving in secret or sentient sympathy with the tendency towards death. That said, the reclusive response to death is very various: from Montaigne's discovery that 'death had a delicate, buoyant quality' that could alter his entire approach to life, through Pope's 'imagination of death', as a presence to haunt and re-enchant his garden with, to Qi Biaojia's ritual suicide, which dramatizes the attendance of death in the garden in a uniquely startling and existential way.[61] Yet what is common to each of these attitudes is the commitment to befriend, and not to run away from, death. For James Hillman, this aligns us with the needs of the soul.[62] To nourish such a deep companionship with death is a principal allegiance of the life of withdrawal and a notable example of its patient exploration of how to accommodate rather than ward off wounding experiences and how to live without maintaining the illusion of binary opposites.

All but one of the gardens referred to in this book were created by people who withdrew from public life for one reason or another – sometimes, initially at least, with mixed feelings – but then found the life of withdrawal they embarked on, in and through their gardens, opened them up to a consoling, or transforming, new version of who they wanted to become and how they needed to dwell in a world that was both beautiful and mysterious, yet also torn in numerous ways that both threatened to defy, while ultimately deepening, that mystery and beauty. What they have left us are their respective visions of the garden as a dwelling-place conceived to partake of, and to reflect, those same conditions of rupture, mystery and beauty; gardens that ensoul and repair and offer the world in some tiny part of it, encouraging us by their example to tend each other also through our way of attending (which, as Malebranche said, is the natural prayer of the soul) to how best to look after a beloved space.[63]

Acknowledgements

This Afterword could not have assumed its present shape without the encouragement and support of Yue Zhuang, whose deep knowledge of and familiarity with the Chinese sources delights and educates me in equal measure. I would also like to pay a more general tribute to Yue for her indispensable role in both conceiving and curating every aspect of this volume. Few co-editors could be as unfailingly generous, insightful and conscientious as Yue.

Notes

1 See Richard C. Goode, 'Loving the World by Leaving it: Connecting Ignatian Silence and Trappist Solitude', in *The Other Journal*, 36, Creation (Fall 2023). Available Online: https://theotherjournal.com/2023/12/loving-the-world-by-leaving-it/, accessed 8 September 2024.

2 For Isaac Luria (1534–72) see Lawrence Fine, *Physician of the Soul, Healer of the Cosmos: Isaac Luria and His Kabbalistic Fellowship* (Stanford, CA: Stanford University Press, 2004). For *tsimtsum*, see Agata Bielik-Robson and Daniel H. Weiss, eds. *Tsimtsum and Modernity: Lurianic Heritage in Modern Philosophy and Theology* (Berlin: De Gruyter, 2021), available as an e-book online: https://doi.org/10.1515/9783110684353

3 See the Introduction for the importance of the tradition of going into hiding in both China and Europe.

4 '[M]ore often than not in Western history it has been the garden, whether real or imaginary, that has provided sanctuary from the frenzy and tumult of history.' Robert Pogue Harrison, *Gardens: An Essay on the Human Condition* (Chicago and London: Chicago University Press, 2008), ix. In East Asia and Iran, the reputation of the garden in this respect is even more renowned than it is in the West.

5 More correctly *tikkun olam* or the 'repairing of the world'. Under Isaac Luria this term, initially used in legal contexts (orientated towards maintaining social cohesion), was given the new mystical meaning of returning the sparks of Divine Light (scattered and buried in the catastrophic creation process of the 'breaking of the vessels' *shevirat ha-Kelim*) to their Divine source.

6 Rosenzweig's notion of *Beseelung* is discussed by Eric Santner in his book *On Creaturely Life: Rilke/Benjamin/Sebald* (Chicago: University of Chicago Press, 2006), 207.

7 The psychologist James Hillman expands the notion of a 'psychic vessel' to embrace the *anima mundi* or World Soul. In that case, 'ensoulment' is not a one-way projection of our idea of soul onto the world, for the world itself is already possessed of soul which actively communicates with us. Ensoulment on this reading is a reciprocal relation. A garden might then be imagined as the vessel where the individual and the World Soul come together in the making of a unique place. See James Hillman, '*Anima Mundi*: The Return of the Soul to the World', in *The Thought of the Heart and the Soul of the World* (Dallas, TX: Spring Publications, Inc., 1981), 91–106.

8 In China, the earliest semi-legendary recluses such as the brothers Po Yi and Shu Ch'i (Boyi and Shuqi), living *c.* 1046 BCE, and Hsü Yu (Xu You) *c.* 2356–2255 BCE and Chieh Chih-t'ui (Jie Zhitui) fl. seventh century BCE, were

venerated by Confucius and others for their uncompromising moral stance in refusing to become corrupted by the offices of power. The two brothers starved to death in the mountains of Shouyang, while Jie Zhitui was burned to death by a fire intended to draw him out of his forest retreat. During the Eastern Han dynasty (25–220) principled withdrawal to a garden became for the first time a possible way of life for scholars who wished to dissociate themselves from a given political regime, inspired by a combination of Confucian, Daoist and Buddhist teachings. Among the surviving written accounts of such a life are Zhang Heng's (78–139) 'Rhapsody on Returning to Fields' *(Gui tian fu)* and Zhongchang Tong's (179–220) 'On Happiness' (*Le zhi lun*). For a historical survey of reclusion in China, see Li Chi, 'The Changing Concept of the Recluse in Chinese Literature', in *Harvard Journal of Asiatic Studies*, 24 (1962–3): 234–47, and, more recently, Aat Emile Vervoorn, *Men of the Cliffs and Caves: The Development of the Chinese Eremitic Tradition to the End of the Han Dynasty* (Hong Kong: Chinese University Press, 1990) [for a Chinese translation, see Aat Emile Vervoorn, *Yan Xue Zhi Shi: Zhongguo Zao Qi Yin Yi Chuan Tong* 岩穴之士：中国早期隐逸传统 (*Men of the Cliffs and Caves: The Development of the Chinese Eremitic Tradition to the Han Dynasty*) trans. Keqian Xu (Ji'nan: Shandong huabao chubanshe, 2009)]; Alan J. Berkowitz, *Patterns of Disengagement: The Practice and Portrayal of Reclusion in Early Medieval China* (Redwood City, CA: Stanford University Press, 2000). For Eastern Han gardens, see Wang Yi, *A Cultural History of Chinese Gardens* [in Chinese], 2nd ed. (Shanghai: Shanghai renmin chubanshe, 2014 [1990]), 75, 72. Within this overall picture, garden retreats, following a tripartite distinction made by the Tang poet Bai Juyi (772–846), are best characterized as belonging to the grade of 'middle reclusion' (*zhong yin*), a term which itself includes a variety of options: so Wang Xianchen and Qi Biaojia, the owners of the two Ming gardens discussed in this volume, developed their gardens, while still quite young, after definitively leaving government service, whereas Bai Juyi was able to combine his style of garden reclusion while remaining in service as a government secretary. 'Middle *yin*' occupied a position poised between 'Great *yin*', on the one hand, referring to those who remained hidden recluses at court or in the midst of city life, and 'Small *yin*', which referred to those who took to the hills. According to Bai Juyi, the former path is too turbulent, while the latter is too solitary. The garden retreat provided a perfect mean.

9 Withdrawal to a garden, even when a kind of exile, always has a voluntary component insofar as it provides an opportunity for honourable self-cultivation even in those, like Derek Jarman, for whom, without the intervention of adverse circumstances, the life of withdrawal goes against the grain.

10 David Cooper (Chapter 6 in this volume) also addresses this question and answers it in broadly similar terms.

11 '[S]o much of the imaginary of reclusion is geographical.' See Gina Cogan, 'Reclusion and Spatiality', in *The Princess Nun: Bunchi, Buddhist Reform, and Gender in Early Edo Japan* (Cambridge, MA: Harvard University Asia Center, 2014), 139–65; Robert Rodriguez, 'Hermit dwelling places: cell, hut and cabin', in *The Book of Hermits: A History of Hermits from Antiquity to the Present* (Hermitary Press, 2021), 259–78.

12 Cogan, *The Princess Nun*, 145–6. A similar recognition was longstanding in China, dating back to descriptions of the preferred environments for different kinds of activity surviving from the Zhou dynasty (*c.* 1046–256 BCE), including this portrayal of the ideal living environment for the Confucian scholar given in the *Book of Rites*: 'The scholar may have a house in (only) a *mu* of ground – a (poor) dwelling each of whose (surrounding) walls is (only) ten paces long, with an outer door of thorns and bamboos, and openings in the wall, long and pointed; within, the inner door stopped up by brushwood, and little round windows like the mouth of a jar . . .' See 'The Conduct of the Scholar', in *The Book of Rites*, English translation cited from https://chinesenotes.com/liji/liji041.html. Confucianism thus provided the tradition of reclusion not only with a rationale of principled withdrawal when appropriate, but also with a concern for the most suitable built environment for withdrawal to occur in. Both (characteristically humane) points of view were influential in the establishment of literati garden culture. Nonetheless, in classical Ming literati gardens, such as the Garden of Artless Administration, the setting is a prosperous suburban environment far removed in actuality (if not in association) from the humble dwelling described above in *The Book of Rites*. What is more notable in Suzhou gardens is a clear distinction between the rational (Confucian) architectural layout of the house and the less formal and geometrical layout of the garden, which is more in sympathy with a Daoist preference for a flowing and irregular organization of space.

13 William Wordsworth, *The Prelude: A Parallel Text*, ed. J. C. Maxwell (Harmondsworth, Middlesex: Penguin Books, 1979) (1805) Bk 10, ll. 729–31: 'But in the very world which is the world / Of all of us, the place in which, in the end, / We find our happiness, or not at all.'

14 I borrow this phrase from Alan J. Berkowitz's *Patterns of Disengagement.*

15 Berkowitz, *Patterns of Disengagement,* 7.

16 Among the earliest non-fictional accounts of the phenomenon of modern urban hermits is that of an anonymous writer, possibly Charles Dickens Jr (son of the novelist), who wrote: 'The hermit is one who goes out into the wilderness to live alone . . . and nowhere is it easier to carry out the eremitical plan than in the wilderness of a great city such as London.' See the article 'Hermits Ancient and Modern' in *All The Year Round: A Weekly Journal*, 74 (24 March 1894): 282–5. For contemporary gardens made by homeless people in the city, see Diana Belmori and Margaret Morton, *Transitory Gardens, Uprooted Lives* (New Haven, CT: Yale University Press, 1995); Robert Harrison, 'Homeless Gardens', in *The Re-Enchantment of the World: Secular Magic in a Rational Age* ed. Joshua Landy and Michael Saler(Redwood City, CA: Stanford University Press, 2009), 72–80.

17 In a letter to Coleridge from Nordhausen, Germany, dated 27 February 1799, Wordsworth distinguished between 'transitory manners reflecting the wearisome unintelligible obliquities of city life [and] manners connected with the permanent objects of nature and partaking of the simplicity of those objects'. *The Letters of William and Dorothy Wordsworth: The Early Years 1787–1805* arr. and ed. Ernest de Selincourt, 2nd ed, rev. Chester L. Shaver (Oxford: Clarendon Press, 1967), 255.

18 For the gradual taming of the setting of reclusion in China during the Han dynasty, see Vervoorn, *Men of the Cliffs and Caves*.

19 The first quotation is from T'ao Ch'ien's (Tao Yuanming, 365–427) famous poem 'Home Again Among Gardens and Fields', a clarion call for the return to a life of retirement in the countryside, in *The Selected Poems of T'ao Ch'ien*, trans. David Hinton (Port Townsend, WA: Copper Canyon Press, 1993). The twelfth-century poet Wang Ch'i (Wang Qi) speaks of being 'secure from the world's polluting dust, / content to dwell in a thatched hut next to a bamboo fence'. See 'The Plum', in *Poems of the Masters: China's Classic Anthology of T'ang and Sung Dynasty Verse*, trans. Red Pine (Port Townsend, WA: Copper Canyon Press, 2003), 510.

20 Confucius famously recommended withdrawal from public service only when 'evil prevails' or when 'the age is disordered'. See, for instance, *Analects* 8.13: 'If the Way is being realized in the world then show yourself; if it is not, then go into reclusion', in Confucius, *Analects with Selections from Traditional Commentaries*, trans. Edward Slingerland (Indianapolis/Cambridge: Hackett Publishing Company, Inc., 2003), 82.The Daoist and Buddhist case for withdrawal was often less conditional on the character of a specific regime (whether corrupt or enlightened) reflecting a wider disillusionment with the human world of affairs more generally. This pervasive scepticism about the virtue of the political realm per se – the sense that the world of men was always 'torn' – gave rise to various ambiguous forms of reclusion as the scholar tried to reconcile maintaining a career with his conscience of fulfilling the ideals of the life of withdrawal. Already in the Han period, the scholar-official Tung-fang so (Dongfang Shuo) (*c.* 160–*c.* 93 BCE), who believed in practising 'retirement at court' (*ch'ao-yin*) and 'retirement in the city' (*shih-yin*), gave this advice to his son in a poem: 'Be a recluse, yet play man of the world; / We are all out of joint with our times.' Quoted in Li Chi, *The Changing Concept of the Recluse*, 242.

21 Li Chi, *The Changing Concept of the Recluse*, 247.

22 For this phrase of Seneca's, see also David Cooper, Chapter 6 this volume.

23 The near legendary status of Tao Yuanming is well attested. He is universally quoted by China's major poets. The mere mention of his name is enough to conjure up the image of the ideal life of virtuous withdrawal lived in harmony with nature, an image Tao himself deliberately promoted in his poetic (auto) biography 'Master of the Five Willows'. There are complete translations of his poems by James Robert Hightower, *The Poetry of Tao Ch'ien* (Oxford: Clarendon Press, 1970) and A. R. Davis, *T'ao Yuan-ming: His Works and Their Meaning* (Cambridge: Cambridge University Press, 1984). Critics based in the West have been less willing than their Chinese counterparts to subscribe to the image of a transparent, spontaneous poet, instead discovering fissures and disturbances beneath the surface serenity and 'seamless selfhood of Master Five Willows'. See Xiaofei Tian, 'Tao Yuanming's Poetics of Awkwardness', in *Companion to World Literature*, ed, Ken Seigneurie (Hoboken, NJ: John Wiley & Sons, Inc., 2019), also available online: https://doi.org/10.1002/9781118635193.ctwl0043; Stephen Owen, 'The Self's Perfect Mirror: Poetry as Autobiography', in *The Vitality of the Lyric Voice: Shih Poetry from Late Han to T'ang*, ed.

Shuen-fu Lin and Stephen Owen (Princeton, NJ: Princeton University Press, 1986), 71–102.

24 See Tian, 'Tao Yuanming's Poetics', 10–11. So, for instance, the celebrated 'Return to Gardens and Fields', I, in one version, instead of ending on a note of serene confirmation 'Now again I return to the natural way of life' has the more hesitant and unsure 'How can I return to the natural way of life?' (translations Tian's).

25 Stendhal famously wrote in *De l'Amour* (1822), chapter XVII, that 'la beauté n'est que la promesse du bonheur' (beauty is only the promise of happiness). The essay, in both French and English, is available in full at gutenberg.org.

26 See Nietzsche's remark made in a note in spring 1888: 'A very popular error: having the courage of one's convictions; rather it is a matter of having the courage for an attack on one's convictions!!!' (*Sehr populärer Irrthum: den Muth zu seiner Überzeugung haben -? Aber den Muth zum Angriff auf seine auf seine Überseuzung haben!!!*) See nietzschesource.org Digital Critical Edition (eKGWB) *Posthumous Fragments* NF-1888, 14[159].

27 See Eric Brown, 'Contemplative Withdrawal in the Hellenistic Age', *Philosophical Studies: An International Journal for Philosophy in the Analytic Tradition*, 137, no. 1 (January 2008): 79–89.

28 See Tom Clucas, *Romantic Reclusion in the Works of Cowper and Wordsworth*, PhD dissertation (Christ Church, University of Oxford, 2014), 35–51.

29 These quotations are taken from the final volume of Stephen Owen's complete translation of Du Fu's poems (with Chinese texts included), *The Poetry of Du Fu*, vols 1–6, trans. and ed. Stephen Owen (Boston/Berlin: Walter de Gruyter Inc., 2016), also available through open access online. References are to Vol. 6, Book 22.69 'Upstairs in the Tower', 22.73 'Respectfully sent to Counselor Lu Ju', and 23.49 'Lying on My Sickbed in the Boat with a "Wind Illness" Writing my Feelings', pp. 111, 119 and 229 respectively. In another poem, 'Setting Out Early' (Vol. 6, Book 22.47, 69–71) Du Fu laments that all his life he has been torn between following 'Wild beans and ferns, starving (like Bo Yi and Shu Qi) on Shouyang' and 'food and horses, provisioned in a series of employments' 'but I'm afraid I have failed on either count'. Even during periods of relative calm, as in the years 760–5, when he was mainly settled into village life near Chengdu in Western China, or when he lived at Kuizhou, at the entrance to the Three Gorges in 766–8, his 'restless and frantic' temperament (22.47), regretting his exile from the capital and distressed by the wretched state of the people as well as his own inability to provide adequately for his family, prevented his 'reclusive expansive heart' (22.58) which 'long had been a Xie Lingyun (385–433, the earlier renowned poet of 'mountains and rivers') accustomed to seeking secluded places' (22.62) from finding lasting equanimity. Examples could be multiplied. For a similar, if less harrowing, account of the tension between a life of service and a life of reclusion in the case of the poet Bai Juyi, see Wang Yi, 'The conflict between service and reclusion and the development of the culture of reclusion', in *A Cultural History of Chinese Gardens*, 210–62.

30 A recent forlorn example is provided by the case of Manfred Gnädinger, a young German who in 1962 built himself a cabin without running water or

electricity on the shore of a wild stretch of the 'Coast of Death' (*Costa da Morte*) at Camelle, Galicia in northwest Spain in order to lead a life of simplicity and solitude, devoted to meditation and art and care for his environment. Over forty years, he crafted a unique sculpture garden and museum from stones and objects he found on the beach, gradually winning the love and respect of the villagers, who knew him simply as 'Man'. Then, in November 2002, the *Prestige* oil tanker broke up off the coast, causing Spain's worst-ever environmental disaster. 'Man' woke up to find his life work and the environment he loved destroyed. In profound sorrow, he retreated to his cabin, locking himself in and pinning a 'Keep Out' sign on the door. A month later, he was found inside dead.

31 See G. K. Chesterton (1935), 'The Case for Hermits', in *The Well and the Shallows* (San Francisco, CA: Ignatius Press, 2007).

32 Henry D. Thoreau, *Walden; or Life in the Woods*, ch. 5 on 'Solitude', para 4. Available online: https://www.gutenberg.org/ebooks/205, accessed 8 September 2024.

33 Dickinson, Poem # 1096: 'Several of Nature's People / I know, and they know me / I feel for them a transport / Of Cordiality', in *The Poems of Emily Dickinson* ed. R. W. Franklin (Cambridge MA: Harvard University Press, 2005).

34 As perceived by Elzéard Bouffier, the hermit protagonist of Jean Giono's *The Man Who Planted Trees* (1953), trans. Barbara May (London: Harvill Press, 1995).

35 'Wordsworth', in Walter Pater (1889), *Appreciations, with an Essay on Style*. Available online through gutenberg.org, 46–7.

36 Wordsworth, *Ode: Intimations of Immortality from Recollections of Early Childhood*, l. 182, in William Wordsworth, *The Poems*, vol. 1, ed. John O. Hayden (Harmondsworth, Middlesex: Penguin Books, 1977).

37 *The Prelude* (1805) Book 1, ll. 425–6, 20. In context, the boy Wordsworth has stolen a boat and rowed out into the lake of Ullswater one evening 'when the moon was up', becoming agitated by the growing stature of a 'huge cliff' which 'as if with voluntary power instinct, Upreared its head' and 'rose up between me and the stars' and 'like a living thing, / strode after me'. The experience provoked 'grave and serious' introspections which haunted the boy for days afterwards (ll. 420–7):

> in my thoughts
> There was a darkness, call it solitude
> Or blank desertion. No familiar shapes
> Of hourly objects, images of trees,
> Of sea or sky, no colours of green fields;
> But huge and mighty forms, that do not live
> Like living men, moved slowly through my mind
> By day, and were the trouble of my dreams.

38 Compare, for example, these two Neo-Confucianist comments from the Song Dynasty: 'The humane person forms one body with all things comprehensively', Cheng Hao (1032–85) and: 'The humane person regards Heaven and earth and all things as one body', Cheng Yi (1033–1107).

39 For a discussion of East Asian rock worship, with its particular resonance in the Zen garden, see François Berthier, *Reading Zen in the Rocks: The Japanese Dry Landscape Garden*, trans. and with a Philosophical Essay by Graham Parkes (Chicago: University of Chicago Press, 2005). See also John Hay, 'The Rock and Chinese Art' in *Orientations*, 16 (1985): 16–32; Graham Parkes, 'The Role of Stone in the Chinese Rock Garden', in *The Bloomsbury Research Handbook of Chinese Aesthetics and Philosophy of Art*, ed. Marcello Ghilardi and Hans-Georg Moeller (London: Bloomsbury Academic, 2021), 277–300. Available online: http://dx.doi.org/10.5040/9781350129795.ch-014, accessed 5 March 2024. The ground plane of the Zen garden, before it represented anything, was simply an empty sacred space, with a provenance dating back to the consecrated clearing (*yuniwa*) associated with a Shinto shrine.

40 Shakespeare's song 'Under the Greenwood Tree' in *As You Like It* and Du Fu's nostalgia for 'being within hearing of the gibbon's cry' are characteristic cultural expressions of the longing to exchange the natural world for the human one.

41 John Donne (1624), 'No man is an Iland, intire of itselfe; every man is a peece of the *continent*, a part of the *maine* . . . any mans *death* diminishes *me*, because I am involved in *Mankinde*; And therefore never send to know for whom the *bell* tolls; It tolls for *thee*', in *Devotions upon Emergent Occasions, and seuerall steps in my Sicknes*, XVII. Nunc lento sonitu dicunt, Morieris. *Now this Bell tolling softly for another, saies to me, Thou must die.* XVII Meditation, in *John Donne Selected Prose*, chosen by Evelyn Simpson and ed. Helen Gardner and Timothy Healy (Oxford: Clarendon Press, 1967), 100–1.

42 See also the examples of Manfred Gnädinger (note 30, above) and Henry Darger (1892–1973), an artist and novelist who lived in complete seclusion in a single room in a Chicago tenement, never showing his work to anyone during his lifetime. His paintings were discovered by accident after his death and are now represented in many of the world's leading museums. I call these cases 'pure' since they exemplify a solitary artistic vocation which is compelled to fulfil itself without any need to communicate to, or seek the approval of, others. The calling is sufficient to itself. In such rare instances, the following, otherwise plausible, insight of Winnicott is belied: 'In the artist . . . one can detect an inherent dilemma' owing to 'the co-existence of two trends, the urgent need to communicate and the still more urgent need not to be found.' See Donald W. Winnicott, 'Communicating and Not Communicating Leading to a Study of Certain Opposites' (1962, 1965), in *The Collected Works of D. W. Winnicott*, vol. 6, 1960–3, ed. Lesley Caldwell and Helen Taylor Robinson (Oxford: Oxford Academic 2016), 438. Available online: https://doi.org/10.1093/med:psych/9780190271381.003.0073, accessed 9 September 2024.

43 Jean-Jacques Rousseau's *The Social Contract* (*Du Contrat Social*) was first published in 1762.

44 In Pali these three types of withdrawal (*viveka*) are known as *kaya-viveka*, *citta-viveka* and *upadhi-viveka*.

45 Thomas Merton, 'Notes on the Philosophy of Solitude', in *Disputed Questions* (New York: Farrar, Straus and Cudahy, 1960).

46 See Elizabeth J. Harris, 'Detachment and Compassion in Early Buddhism', *Access to Insight (BCBS Edition)*, 30 November 2013. http://www.accesstoinsight.org/lib/authors/harris/bl141.html, accessed 8 September 2024.

47 *Lines Composed a Few Miles above Tintern Abbey, on Revisiting the Banks of the Wye during a Tour (July 13, 1798)*, ll. 41, 49, in *The Poems*, vol. 1.

48 Daguerre's park at Bry, the subject of Kasie Alt's Chapter 2 in this volume, is an early example of a landscape designed, at least in part, for public access. It remained, however, in private ownership and was privately funded. A few years earlier (1847) Joseph Paxton's Birkenhead Park had become the first publicly funded civic park in the world.

49 See Maynard Mack, *The Garden and the City: Retirement and Politics in the Later Poetry of Pope 1731–1743* (Toronto: University of Toronto Press, 1969), 232. For Pope and the 'place to stand', see also Yue Zhuang, Chapter 9 in this volume.

50 'But there are another sort of people who seem design'd for solitude, such I mean as have more to hide than to show: As for my own part I am one of those . . .' Alexander Pope, Letter to Mr Steele of June 18, 1712, in *The Correspondence of Alexander Pope*, vol. 1, 1704–18, ed. George Sherburn (Oxford: Oxford University Press, 1956), published online October 2017.

51 Mack, *The Garden and The City*, 107.

52 Pope's principled withdrawal and his 'place to stand' led to two other remarkable poets' gardens in the UK. His elegiac stance was pursued more consistently by another eighteenth-century poet, William Shenstone (1714–63) in his Virgilian Arcadia at The Leasowes, while his critique of contemporary culture was brought into the modern era by Ian Hamilton Finlay (1925–2006) at Little Sparta. Finlay, however, remained an admirer of Shenstone, too, and what Mack said of Pope might be said equally of him: 'the retirement tradition was always under the surface of Pope's consciousness and may at any time break out' (Mack, *The Garden and The City*, 113).

53 Virgil, *Georgics* II, 502.

54 See note 42.

55 Stendhal famously ends his novel *La Chartreuse de Parme* (The Charterhouse of Parma) with a dedication 'to the happy few' (in English), a phrase he may have borrowed from either Oliver Goldsmith (*The Vicar of Wakefield*, ch. 2) or Shakespeare (*Henry V*, Act IV Scene iii, 60).

56 'How safe, methinks, and strong, behind / These trees have I encamped my mind.' Andrew Marvell (1651), 'Upon Appleton House', in *The Poems and Letters of Andrew Marvell*, vol. I, Poems, lines 601–2, ed. H. M. Margoliouth (Oxford: Clarendon Press, 1971).

57 For the failing of innocence in this respect, see Ian Kidd, Chapter 7 in this volume. In poetry, the image of the virtuous countryman living in retirement crystallized in, and became passed down through, the trope of the *Beatus vir* (happy man) as formulated, for instance, by the Roman poets Horace (65–27 BCE) *Epode* 2, and Claudian (d. 404 CE) *De Sene Veronensis Suburbium numquam egressus est* (Of an old Man of Verona who never left his home). The happy man is a freeholder who inherits a farm from his father and finds

contentment in working the land without ever leaving it. He leads a virtuous and industrious life, sleeping soundly and cherishing his independence, free of all business and financial cares. This life of innocence in the countryside is often supplemented, as in Virgil, by the complementary ideal of the retired philosopher.

58 Abraham Cowley, 'The Garden', in *Essays*, ed. Henry Morley (London: Cassell & Company Ltd, 1893), 103–118. Available online: https://www.gutenberg.org/files/3549/3549-h/3549-h.htm, accessed 8 September 2024. The attention of participants in the workshop which gave rise to this book was drawn to Cowley through an illuminating contribution made by Henry Power, Professor of English at the University of Exeter.

59 Wallace Stevens, 'Sunday Morning' VI, in *The Collected Poems of Wallace Stevens* (London: Faber and Faber Ltd, 1955).

60 'Superior gardens are composed of Glooms and Solitudes and not of plants and trees.' This is one of Finlay's 'Unconnected Sentences on Gardening' in *Ian Hamilton Finlay: Selections*, ed. Alec Finlay (Berkeley, CA: University of California Press, 2012). The 'Unconnected Sentences' are also freely available online. See also from his 'Detached Sentences on Exile': 'A garden, being less a place than a world, is a proper work for an *exile*' and 'Death is the extremest *exile*' (italics Finlay's).

61 For Montaigne, see the website returnofanative.com *Gazing Inward, Michel de Montaigne*. It is perhaps no accident that the qualities of 'delicacy' and 'buoyancy' are detected as pivotal on a website that 'explores the lives of creative people and their relationship to place'. For Pope and Qi Biaojia, see Yue Zhuang (Chapter 9) and Lin Zhu (Chapter 1) in this volume.

62 '[T]he significance soul makes possible, whether in love or religious concern, derives from its special *relation with death*.' James Hillman, *Re-Visioning Psychology* (New York: Harper Perennial, 1992), xvi (italics Hillman's).

63 Nicolas Malebranche (1674), *The Search after Truth*, trans. and ed. Thomas M. Lennon and Paul J. Olscamp (Cambridge: Cambridge University Press, 1997), 529.

INDEX